Genetics and reductionism

With the advent of the Human Genome Project there have been many claims for the genetic origins of complex human behaviors including insanity, criminality, and intelligence. But what does it really mean to call a trait "genetic"? This is the fundamental question that Sahotra Sarkar's book addresses.

The author analyzes the nature of reductionism in classical and molecular genetics. He shows that there are two radically different kinds of reductionist explanation: genetic reduction (as found in classical genetics) and physical reduction (found in molecular genetics). A behavioral trait can only be said to be genetic if genes alone provide the best explanation of its origin, and virtually all complex human behavioral traits fail this test.

This important book clarifies the meaning of the term "genetic," shows how molecular studies have affected genetics, and provides the philosophical background necessary to understand the debates over the Human Genome Project.

It will be of particular interest to professionals and students in the philosophy of science, the history of science, and the social studies of science, medicine, and technology.

Sahotra Sarkar is a Fellow of the Wissenschaftskolleg zu Berlin.

CAMBRIDGE STUDIES IN PHILOSOPHY AND BIOLOGY

General Editor
Michael Ruse *University of Guelph*

Advisory Board
Michael Donoghue *Harvard University*
Jean Gayon *Université de Paris 7*
Jonathan Hodge *University of Leeds*
Jane Maienschein *Arizona State University*
Jesus Mosterin *University of Barcelona*
Elliott Sober *University of Wisconsin*

This major series publishes the very best work in the philosophy of biology. Nonsectarian in character, the series extends across the broadest range of topics: evolutionary theory, population genetics, molecular biology, ecology, human biology, systematics, and more. A special welcome is given to contributions treating significant advances in biological theory and practice, such as those emerging from the Human Genome Project. At the same time, due emphasis is given to the historical context of the subject, and there is an important place for projects that support philosophical claims with historical case studies.

Books in the series are genuinely interdisciplinary, aimed at a broad cross-section of philosophers and biologists, as well as interested scholars in related disciplines. They include specialist monographs, collaborative volumes, and – in a few instances – selected papers by a single author.

Alfred I. Tauber: *The Immune Self: Theory or Metaphor?*
Elliott Sober: *From a Biological Point of View*
Robert Brandon: *Concepts and Methods in Evolutionary Biology*
Peter Godfrey-Smith: *Complexity and the Function of Mind in Nature*
William A. Rottschaefer: *The Biology and Psychology of Moral Agency*

Genetics and reductionism

SAHOTRA SARKAR

CAMBRIDGE
UNIVERSITY PRESS

PUBLISHED BY THE PRESS SYNDICATE OF THE UNIVERSITY OF CAMBRIDGE
The Pitt Building, Trumpington Street, Cambridge CB2 1RP, United Kingdom

CAMBRIDGE UNIVERSITY PRESS
The Edinburgh Building, Cambridge CB2 2RU, UK   http://www.cup.cam.ac.uk
40 West 20th Street, New York, NY 10011-4211, USA   http://www.cup.org
10 Stamford Road, Oakleigh, Melbourne 3166, Australia

First published 1998

Printed in the United States of America

Typeset in Times Roman 10/12 pt. in LaTeX 2_ε [TB]

A catalog record for this book is available from the British Library

Library of Congress Cataloging-in-Publication Data
Sarkar, Sahotra.
Genetics and reductionism / Sahotra Sarkar.
p.   cm.
ISBN 0-521-63146-7 (hbk.). – ISBN 0-521-63713-9 (pbk.)
1. Human genetics – Philosophy.   2. Reductionism.   3. Genetic
psychology.   4. Human Genome Project.   I. Title
QH431.S313   1999
599.93′5′01 – dc21   98-20460
                                                              CIP

0 521 63146 7 hardback
0 521 63713 9 paperback

*To my parents, Niladri and Malabika Sarkar, for my genes and, even more
so, for my formative environment*

Contents

Contents

Acknowledgments

This work was partly supported by a grant from the National Institutes of Health (No. HG 00912-02/3), a Fellowship from the Dibner Institute at MIT, a Fellowship from the Edelstein Center at the Hebrew University of Jerusalem, and a Fellowship from the Wissenschaftskolleg zu Berlin. This book was largely written during the course of two years spent at McGill University in the Philosophy Department. Colleagues and students there provided ample encouragement. The students – Robin Copp, Eric Cytrynbaum, Hyung Chan Kim, Michael Pressau, and Christopher Studnicki-Gizbert – suffered through many drafts of this book as the various chapters were written and rewritten. The manuscript was finally revised for publication while the author was a Fellow of the Wissenschaftskolleg zu Berlin. The facilities there left nothing to be desired for academic work. Though the staff of that institution did not actually write this book, their help alone made its final completion possible.

For reading and commenting on the entire manuscript, thanks are due to David Hull, Kenneth Schaffner, William Wimsatt, and an anonymous reviewer for Cambridge University Press; to Jordi Cat and Gregg Jaeger for reading and commenting on Chapters 1, 2, and 3; to Storrs McCall for Chapter 2; to Raphael Falk for Chapters 4 and 6; to Ned Block and Evelyn Fox Keller for Chapter 4; and to Manal Swairjo Jaeger and Denis Thieffry for Chapter 6. Comments made by Richard Hudson on the analysis presented in Chapter 4 were also very useful. Thanks are due to Jonathan Beckwith, Werner Callebaut, Raphael Falk, Scott Gilbert, Richard Lewontin, Kenneth Schaffner, Abner Shimony, John Stachel, and William Wimsatt for many illuminating discussions about reductionism over the last decade. Different chapters of this book were presented as seminars or colloquia to the Departments of Philosophy at the University of New Mexico, Brandeis University, the University of Western Ontario, and the University of California at Irvine; the Department of the Social Studies of Medicine at McGill University;

Acknowledgments

the Department of Biology at McGill University; the Population Genetics Laboratory at Harvard University; and the Belgian Society for Logic and Philosophy of Science. Comments and criticism by members of the various audiences were uniformly helpful. Thanks are also due to Terence Moore and Michael Ruse for encouraging this project. Greger Jaeger checked the references. Needless to say, none of the individuals mentioned above should be blamed for any of the errors that this book still contains. Those errors should, instead, be attributed to Stephen Menn, who rescued and read many chapter drafts from the recycling bins at McGill's philosophy department. Most of this book was initially composed at the Shed Cafe in Montréal, a place particularly conducive to uncritical philosophical reflection.

1

Introduction

When Nolan Ryan was asked to explain his remarkable longevity in the major leagues, he attributed it mostly to genetics.[1] There is very little doubt that the genetic explanation of human behavior has returned with a vengeance, not only in scientific and semiscientific circles, but in the popular imagination as well. This renewed faith in genetics has emerged hand-in-hand with a new-found popular deification of DNA. New Age bookstores in San Francisco have begun offering music based on the frequency spectra of DNA molecules (raised by 35 octaves to be brought into the human audible range).[2] A New York company has developed wind chimes from the same basis. In Paris, a biotechnology company has been offering *Le Biopen* with DNA-laced ink. The fifty-base sequence of DNA is specific to each bottle of ink and is kept secret by the company so that forgeries may be detected. A pen and a bottle of ink cost about $10,000, and several European and Japanese companies are said to be interested. To top it all, a California company plans to sell pieces of DNA – possibly encoding genes – cloned from pop idols including rock musicians and movie stars.

Though hereditarian thinking has consistently been part of the Western intellectual tradition for over two centuries, genetic explanations of human behavior last enjoyed this level of hegemony in the 1920s and 1930s, at the height of eugenic enthusiasm in the United States and Europe. Eugenics – and its attendant research programs – then disappeared, largely due to the Nazi abuse of biology, though, at least arguably, eugenics was never scientifically refuted.[3] Human genetics, which had struggled under the yoke of eugenics in the 1930s, finally emerged as a relatively depoliticized discipline in the 1950s.[4] While it paid welcome attention to the genetic origin of biochemically characterized diseases (the "inborn errors of metabolism," to use Garrod's [1909] memorable phrase) including sickle cell disease and phenylketonuria (PKU), traditional human genetics (during the 1950–80 period) was largely silent about those complex behavioral patterns such as

1

insanity, criminality, or vagrantism that had so infatuated the eugenicists.[5] Human geneticists did not usually claim that social problems – poverty, crime, or even the unequal distribution of wealth – had genetic origins, or should be addressed through genetic interventions.

Such modesty has disappeared. Of late, genes have been directly invoked to explain, among many other complex human behavioral patterns, male sexual orientation, schizophrenia, alcoholism, autism, reading disability, bipolar affective disorder (more popularly known as manic depression), neuroticism, adolescent vocational interests, spatial and verbal reasoning, and even alleged differences in intelligence (as measured by the variability of IQ scores).[6] There have been conferences dedicated to exploring the possible genetic basis for criminal behavior.[7] At one point, an editorial in as respected a journal as *Science* suggested that genetics could be used to solve problems such as that of a "crazed gunman terrorizing hostages in a bar at Berkeley."[8]

Between the last and present incarnations of this sort of "geneticism" – to coin a term – much has changed. The well-known "triumphs" of molecular biology in the 1950s and 1960s – the double helix model of DNA structure, the operon model of gene regulation, the allostery model of protein interactions, the genetic "code," etc. – have transformed biology to an extent only comparable to the rise of evolutionary thinking during the last century. Genetics, the locus of much of the research in molecular biology, stands particularly transformed. Whereas the imputation of human behavioral patterns to genes in the 1920s relied on little more than the semi-intuitive analysis of the distribution of traits up a pedigree, contemporary geneticists not only map traits to specific loci on chromosomes (which is hardly new), but then proceed to sequence segments of DNA and attempt to identify and investigate the putative proteins through which a gene may be expressed. Geneticists promise to provide a complete catalog of human genes within the next decade; some go even further and promise what they call a complete "human genome sequence." Genetic explanations are now offered with a qualitatively new degree of precision or, at least, the appearance of such precision.

While the new studies reporting alleged links between genes and complex human behavioral patterns have routinely been hailed as the birth of a "new genetics," they have not gone unchallenged.[9] For instance, in the case of bipolar affective disorder, continued investigation of the same pedigrees that had suggested linkage (that is, roughly, a tendency for the disorder to be inherited along with a known gene) undermined the original claims.[10] Similarly, in the case of schizophrenia the original findings indicating linkage to a gene could not be replicated in some new pedigrees. The case of

2

alcoholism has been even more confusing. It had long been suggested that alcoholism was more likely to be observed in family members of alcoholics than in the general population. A report from 1990 suggested that there was an association between alcoholism and a particular gene on chromosome 11 – the A1 allele for the dopamine D2 receptor (DRD2) gene.[11] Subsequent studies have sometimes verified that report and, just as often, denied it. A relatively recent meta-analysis comes out very cautiously in favor of the increase.[12] It is generally (though by no means universally) agreed that systematic methodological flaws continue to plague the studies that claim to indicate a heritable basis for intelligence differences.[13]

Only two things appear certain: (i) as the Human Genome Project (HGP) – the attempt to identify and map all human genes and (possibly) to provide an entire sequence for a "human genome" – proceeds, there will be increasingly more claims of the genetic origin of such complex behavioral (and other) human features; (ii) the controversies over these will not disappear in the short run. The reason for this is that most of these controversies are ultimately generated not by inadequate data, as in the case of bipolar affective disorder, which can potentially be remedied in a (relatively) straightforward fashion. Rather, the controversies usually arise from more difficult methodological and interpretive issues. Does the estimation of "heritability," a concept that will be scrutinized in Chapter 4, truly show a genetic basis for anything? Does linkage analysis (described in Chapter 5) establish anything more than a correlation? Does it even make sense to attribute some features to the genes if they interact significantly with nongenetic (environmental) factors? Are techniques such as the analysis of variance sensitive to such interactions? These are the sort of questions that will occupy this book, though only when (and that is quite often) they are related to the issue of reductionism. The rest of this Introduction will indicate the general strategy of this book and – though only in a rather pedestrian fashion – note those points of philosophy or biology that usually receive inadequate attention in standard textbook treatments.

1.1. WHAT'S "GENETIC"?

In both technical and popular accounts of these disputes, the question that has almost universally been debated is whether a particular behavioral or structural feature of an organism is genetic. In many ways, the dispute over what is genetic is a contemporary rendition of the traditional nature-nurture dispute.[14] What is genetic is to be contrasted with what is environmental (which, in this context, is usually taken to include everything that is not genetic). It is not only routinely taken for granted that what is genetic can be categorically distinguished from what is environmental, but it is often even assumed that a definite answer – one way or the other – must be forthcoming for every feature.

It has usually gone unnoticed that, before whether some feature is genetic can be determined, what must be agreed upon is what it means to call something "genetic." No technical result defines that term and, unfortunately, there is no easy way to define it. Resort to the usual philosophical strategies of definition – for instance, the introduction of genes as necessary or sufficient conditions for the features, or an appeal to genetic causation – is of little help. If it be suggested that a feature should be called "genetic" if and only if some gene (or set of genes) is necessary for its genesis, virtually all organismic features would turn out to be genetic because of the simple fact that many genes that act during the earliest stages of embryonic development are always necessary for the continued development of an embryo. If it be suggested that a feature is "genetic" if and only if some gene (or set of genes) is sufficient for its genesis, nothing would turn out to be genetic because any gene can act only in suitable (biochemical or developmental) environments. Even if this problem is somehow avoided, this putative definition has an additional unsatisfactory feature: it does not permit any late-developing feature to be uniquely attributed to a gene that would – in ordinary usage – be held responsible for it. Because of the importance of the genes acting early in embryonic development, neither polydactyly nor the sickle-cell trait could be uniquely attributed to their well-known genes. Even among only other genes, these do not suffice for the genesis of these traits since the genes acting early in development are necessary for all traits. Falling back on both necessary and sufficient conditions does not change this situation in any way.

Another philosophically motivated potential strategy of definition is to require that some feature be called "genetic" if and only if some gene (or set of genes) *causes* the genesis of that feature.[15] This, too, is of no help. It simply transfers the difficulty of defining "genetic" to the even more troublesome problem of defining "cause." Moreover, for the reasons already indicated, that every organismic feature requires both a set of necessary genes *and* a set of necessary environmental characteristics, while neither is sufficient by itself, it would appear that any reasonable definition of "cause" would find both genes and some environmental factors to be causes of any organismic feature of interest. There would be no possibility of distinguishing genetic features from those that were not, that is, those whose etiology involved environmental factors as much as, if not more than, any gene. Recall that this was the motivation behind the search for the definition in the first place.

The difficulty of defining "genetic" is severe enough that some philosophers have suggested that the attempt simply be abandoned.[16] However, such extreme pessimism seems to be unwarranted. An attempt will be made in this book (in Chapter 7, § 7.2) to define "genetic," but the route followed will be circuitous. First, the nature of genetic explanation – when genes serve as the best explanation of an organismic feature – will be analyzed. (This is roughly what will be meant by "reduction" to genetics in this book.) The result of that analysis will then be used to formulate a proposal for the use of "genetic." This proposal will involve some conventional elements and will not allow the recovery of all ordinary uses of "genetic." This should come as no surprise for at least two reasons: (i) if there is a way for all these uses to be captured by any definition, it is unlikely that the terminological controversy over "genetic" would have arisen in the first place; and (ii) almost all philosophical explications that attempt to systematize a term by regulation through convention involve some loss of original connotations, and "genetic" is no exception in this regard. In any case, what should be emphasized here is that the main purpose of this book remains the analysis of genetical methodology and, in particular, genetic explanation, that is, reduction. Whether the explication of "genetic" is also successful is only of secondary importance.

1.2.   ALLELES, LOCI, AND TRAITS – A NOTE
ON BIOLOGICAL TERMINOLOGY

To avoid unnecessary potential confusion, some genetic terminology that will later be used in this book will be clarified in this section. As has been pointed out many times before, the most basic terms of genetics, including "gene," are surprisingly ambiguous. To begin, the "genotype" of an organism is, informally, the set of all its genes. The reason why this characterization is informal is that a gene is an abstract entity that is instantiated in every organism in some of many potential versions called "alleles." Each gene is instantiated at a particular location on a chromosome (or similar gene-instantiating structure) of organisms.[17] That location is called its "locus," which is only defined relative to other loci in classical genetics. Chromosomes instantiating particular genes can occur singly, in pairs, or in sets of higher cardinality in different organisms. Sets of chromosomes (or, often, the chromosomes themselves) that contain the same loci are called "homologous." If an organism contains only single chromosomes, it is called "haploid." If it contains paired homologous chromosomes, it is called "diploid."[18] If the sets of homologous chromosomes have higher cardinality, the organism is said to be "polyploid." In general, the cardinality of such a set is called the "ploidy" of an organism. When the ploidy is greater than 1, a "locus" refers to a gene's location on all of the homologous chromosomes. (One therefore speaks of two alleles at one locus in diploids, etc.)

Conventional Mendelian genetics, which applies to most "higher" animals including humans, assumes diploidy. Throughout this book, unless explicitly indicated otherwise, Mendelian genetics (including, therefore, diploidy) will be assumed.[19] If a diploid organism has two identical alleles at a locus, it is "homozygous" at that locus; otherwise it is "heterozygous." Mendelian genetics, in its contemporary form – sometimes called "classical genetics" – requires (i) that an individual's two homologous chromosomes each come from each of its two parents; (ii) that two alleles on different chromosomes be inherited independently; and (iii) that two alleles on the same chromosome have a probability of being inherited together that is (roughly) inversely proportional to the separation of their loci on the chromosome.[20] That such alleles might not be inherited together is due to a process called "recombination" (a result of homologous chromosomes exchanging parts as the germinal cells are being formed). The probability that two alleles are inherited together is a measure of the "linkage" of the corresponding loci.

In the context of population genetics, which studies the transmission of genes in populations, most of the ambiguity of the term "gene" comes from its use sometimes for "allele" and at other times for "locus." All of population genetics could be formulated without using the term "gene," and using only "allele" and "locus" instead; it would be a clearer subject as a result. Throughout this book, "allele" and "locus" will be used preferentially over "gene" in all population genetic contexts and anywhere else where this additional precision is useful. In molecular genetics, an additional ambiguity of "gene" arises from its use also to refer to an actual segment of DNA that encodes the (abstract) gene through its sequence. This problem will be encountered and broached in Chapter 6.[21]

With these clarifications, an organism's "genotype" is the complete set of all of its alleles at all loci. The alleles are related to many (even if not all) of the structural and behavioral features of organisms. This relation is at least one of partial determination. However, it is seldom a one-to-one correspondence. Because of this, there is an important distinction between an organism's genotype and its physical and behavioral features, which comprise its "phenotype." This distinction, on which much of modern genetics is based, goes back to Mendel. He distinguished between the "factors" (in modern terminology, the alleles), which were responsible for the "characters" exhibited by the peas that he was studying. In Mendel's example, the set of factors comprised what later came to be called the "genotype," the set of characters the "phenotype." However, the distinction does not require that the genotype has any particular structure, for instance, that there are discrete factors responsible for a character, or that these factors obey Mendel's rules. It is much more general and just presupposes two different realms, one being the genotypic or genetic level with entities that are (at least sometimes) directly inherited, and the other being the phenotypic level, the features of which are to some degree a result of features at the genotypic level, and which are not directly inherited.

At this stage, however, two problems arise: (i) which of its physical and behavioral features are a part of an organism's phenotype, and (ii) how is an organism to be cataloged into a set of phenotypic characteristics? For instance, in humans, is body density a phenotypic feature? Or the number of teardrops shed in a lifetime? These two problems are not independent but also not identical. They are related because if one knew how to catalog an organism into a complete set of distinct phenotypic features, then this set would define its phenotype. The problems are, however, not identical, because even if one knew what sort of feature was allowed to be part of an organism's phenotype (solving the first problem), one might still not be able to provide a complete catalog. For instance, suppose that only those features of an organism that can be directly acted upon by natural selection form part of its phenotype. This criterion is clear enough and, in any definite model of selection, it would be reasonably clear what an organism's phenotypic features could be. However, there is an immense variety of possible models of selection and, in practice, it would still be impossible to provide a complete (context-independent) catalog of the organism's phenotypic features.

There is no technical solution to these problems. There are, however, heuristic proposals with some merit. Consider the following two options.

(i) In response to the first problem, it is reasonable to consider as part of its phenotype any feature of an organism that is nontrivially involved in its interactions with its environment. A full defense of this proposal is beyond the scope of this book; suffice it here only to note that, since natural selection generally acts on an organism through the mediation of its environment, this proposal has the merit of providing an immediate method of incorporating evolutionary considerations into a genetic context.

(ii) In response to the second problem, the catalog of phenotypic characteristics can reasonably include any feature of an organism that shows sufficient regularity or constancy in a given context to be recognized as such. Density would qualify (for humans); the number of teardrops probably would not. The justification for this proposal is that the set of features that biologists study continues to expand – after all, no technical result determines the limits of what is "properly biological" – and this proposal allows the introduction of any feature that provides a reasonable prospect for experimental investigation.

Once any proposal of this sort is accepted, the features that are thus identified are usually called "phenotypic traits," or simply "traits." It is quite likely that the two proposals considered here would provide the same answer in many cases, but not all. For instance, there is no reason to believe that humans interact significantly with their environments through their densities (except, perhaps, in populations living along the edges of bodies of water). Thus, proposal (i) would ignore density, whereas proposal (ii) would embrace it as a trait. To avoid discrepancies of this sort, one could either use their conjunction, if one wanted to be restrictive, or their disjunction, if one wanted to be inclusive. The latter is more reasonable for four reasons: (i) it captures what are uncontroversially regarded as traits such as the characters of Mendel's peas, eye color or the sickle-cell trait (in humans); (ii) it is consistent with normal biological research where the choice of some feature to be studied as a trait is arbitrary in the sense that there are no precise rules governing such a choice but, nevertheless, nonarbitrary in the sense that anything devoid of at least some constancy or regularity simply could not be studied; (iii) such a way of characterizing traits makes no claim whatsoever about their origin. In particular, it introduces no implicit assumption about the role of the genes or the environment in the genesis of a trait. Therefore, these are interesting questions that remain to be investigated; and (iv), as has been noted before, the first and, therefore, the inclusive proposal easily allows these (genetical) studies to be conjoined to evolutionary investigations.

The inclusive proposal for defining a "trait" will be used throughout this book. Note that, even though it is quite inclusive, it does allow the exclusion of some features, for instance, the number of teardrops shed by humans through their entire lives. Moreover, it allows interesting questions to be asked. Leave aside the controversial question of the origins of IQ. Is IQ even a trait? Does it have sufficient constancy or regularity? Is there any evidence that it plays a significant role in an individual's interactions with the environment? Does it influence fitness? For those features that can be called traits, it can then be asked whether genes provide the best explanation for their genesis. This book explores the conceptual and methodological problems associated with such explanations. Finally, the problem of defining "genetic" now becomes that of defining a "genetic trait." As noted before, this problem will be explored in Chapter 7, § 7.2.

1.3. EXPLANATION AND REDUCTION

Explanations bridging two different realms (or levels of domains of inquiry) are often called "reductions" by philosophers, especially if the *explananda* (what are to be explained) are routinely from one realm, and the *explanans* (what does the explaining) are routinely at the other[22]. Beyond this rather broad assessment, there is little philosophical consensus about the nature of "reduction," and even this broad assessment is not entirely uncontroversial.[23] In the present context, that is, in explanations in genetics, the two realms are those of the phenotype and the genotype. The explananda are in the phenotypic realm, while the explanans are potentially at the genotypic realm, only "potentially" because there is ample reason to suspect that a genetic explanation will not always be successful.

Philosophical models of reduction usually attempt to characterize *reductionist explanation*; that is, they attempt to capture both explanation and reduction together. This strategy will not be followed here. Rather, a deliberate attempt will be made to keep the issues of reduction and explanation distinct. The motivation for this is the lack of philosophical consensus about the formal or even the substantive nature of explanation (see Chapter 3, § 3.1).[24] As a result, there have been ostensible disputes about the nature of reduction that are, at least partly, disputes about the nature of explanation.[25] Disputes that really are disputes about explanation in general, rather than specifically about the nature of reduction, will be ignored here so that the discussions retain their focus on the subject that is of explicit concern, that is, reduction.

The strategy adopted will be to assume that one is given an explanation, which qualifies as an explanation according to whatever explication that one may prefer, and then attempt to identify those *additional* criteria that the explanation must satisfy in order to be a reduction. There is an obvious danger in adopting this strategy, namely, that these additional criteria may not cohere with some possible explication of explanation. There does not appear to be any a priori way to guard against this danger fully. All that will be tried here is to make sure that the criteria for reduction remain general enough that they are not likely to fall afoul of most variants of the usual explications of explanation and, more explicitly, that they can be used in conjunction with the deductive-nomological and statistical relevance models of explanation. (A special status is granted to these two models simply because they, along with their many variants, appear most likely to win popularity contests among philosophers.)

However, in many of the discussions, putative reductions will be criticized not because they fail to satisfy various criteria for reduction, but simply because of their inadequacy as explanations. These criticisms, too, will be kept as independent of particular models of explanation as possible. Obviously, as philosophical strategy, this is not altogether satisfactory – it does not have the kind of formal precision that (at least analytic) philosophers have come to prefer. Nevertheless, two things can be said in its favor: (i) because of the lack of consensus about the formal or substantive nature of explanation (that has already been mentioned), it would probably be less convincing to argue against the soundness of a putative explanation on the basis of a particular model of explanation than to provide an argument that is less restrictive; and (ii), more importantly, it is doubtful whether any criticism of a particular scientific explanation is of much value – let alone has the potential to be convincing – if it depends entirely on a particular philosophical formalism used to explicate "explanation." In general, the success or failure of an explanation depends on substantive scientific or methodological issues that constrain what might constitute an adequate model of "explanation," rather than the converse.

The thesis that reductions between two realms are always (or, at least, are always likely to be) successful (as explanations) is often called "reductionism."[26] The thesis that all phenotypic phenomena can (always) be reduced to facts at the genotypic level will be called "genetic reductionism." This thesis is to be distinguished from a different reductionist thesis that has also guided many biological research programs during the last 150 years (including some important programs within genetics), namely, that all biological phenomena are to be explained from a physical basis. The latter thesis will be called "physical reductionism." This book is about reduction in genetics in general and will explore both genetic and physical reductionism. Chapters 4 and 5 are exclusively concerned with the former; the latter is taken up in Chapter 6.

1.4. DETERMINISM AND PREDICTION

There are two concepts related to explanation/reduction that are both part of the popular conception of genetics and are often conflated with explanation/reduction: *genetic determinism* and *prediction*. This book is not about either or these concepts, but since they (especially the latter) sometimes become relevant to the issues dealt with in this book, a short discussion of them at this point will (presumably) help decrease any potential confusion that may arise later. Genetic determinism is motivated by the (correct) observation that there are cases where the possession of a particular allele appears to ensure the possession of a particular trait (e.g., some eye colors, or Huntington's disease in humans) at least in sufficiently old individuals. It is easy enough to formulate criteria for "genetic determinism" that attempt to incorporate this observation. Consider four versions of a putative criterion for genetic determinism:[27]

(i)   for any locus, two individuals with the same alleles at that locus will always exhibit an identical corresponding trait;

(ii) for some loci, two individuals with the same alleles at that locus will always exhibit an identical corresponding trait;

(iii) two individuals identical with respect to all alleles at all loci will always be identical with respect to all traits;

(iv) two individuals identical with respect to all alleles at all loci will always be identical with respect to some of their traits.

Trivially (i) is stronger than (ii); that is, (i) implies (ii) and, similarly, (iii) is stronger than (iv). Only a little less trivially, (i) is stronger than (iii) and (ii) is stronger than (iv). More precisely, (i) implies (iii) and (ii) implies (iv), which suffices to explain their relative strengths. What is more important is that there is some interesting biology behind this. The difference between the first and last two pairs of criteria is that only the latter allow more than one locus to be involved in the genesis of a trait; that is why they are weaker (or, easier to satisfy). More specifically, suppose that there are two loci, **A** and **B**. Suppose that a high frequency of individuals with identical alleles at **A**, but not all of them, exhibit some phenotype P. In such a situation, criterion (i) clearly fails, and assuming that **A** is the locus of interest (for whatever reason), (ii) also fails. However, suppose that individuals with identical alleles at both **A** and **B** always display P. Then (iii) and (iv) are satisfied. Thus, in this sense (iii) and (iv) are weaker than (i) and (ii).

Obviously, there are many possible criteria that fall between ((i), (iii)) and ((iii), (iv)): new criteria can be generated by simply requiring different degrees of genetical identity. These will not be pursued here because, strictly speaking, none of these criteria (and, consequently, no intermediate one) is satisfied. Consider (iv), the weakest. Even it cannot be satisfied: one can, for instance, through sufficient nutritional manipulation during early embryonic development, induce identical twins to differ in any chosen trait – if necessary, simply by starving one of the twins to death early enough during development. This may seem to be a trivial objection: after all, it would appear to be removable by simply restricting what is done to the individuals during development, that is, by making a restriction to a set of "normal environments" (assuming that these can be fully specified, which is not easy). The point, however, is that if genetic *determinism* is truly what is at stake, a move to include environmental variables is unwarranted. Perhaps "genetic determinism" should be construed in a much weaker fashion, by simply requiring genes to be the most important factors in the genesis of a trait, ensuring that two individuals with (appropriately) identical genes usually develop some trait to be identical. But this is not determinism. It is no more than invoking genes to explain traits, that is, it is *reductionism* (as construed here).

Genetic determinism is thus a stronger requirement than genetic reductionism: genes may provide the best explanation for the genesis of a trait without determining it entirely by themselves. In fact, as the argument given above shows, genetic determinism is so strong that it is a vacuous doctrine. At a less general level, restricting attention to "interesting" traits such as the controversial behavioral traits that have already been mentioned, even the most ardent advocates of a genetic basis for behavior concede that genetic determinism is untenable due to the obvious influence of nongenetic factors.[28] There will be no more consideration of genetic determinism in this book. It should be emphasized, however, that dropping genetic determinism from consideration does not decrease the importance of the more philosophically and scientifically interesting question of the extent to which genes can be used to explain complex human behaviors and other traits.[29]

As far as definition goes, in genetics and elsewhere, "prediction" is not as complicated as "determinism." Roughly (and no more precision is required here), to predict the behavior of a system is to state what the values of its (relevant) parameters will be at a time $t' > t$ from a knowledge of what values those (or other) parameters have at some time t. Prediction, unlike determinism, thus has an epistemological component (because of the reference to what is known at time t). More importantly, it has a pragmatic component (which, depending on one's philosophical commitments, may or may not be regarded as part of its epistemological component): in practice, the parameters of a system can usually only be (empirically) known and (computationally) used at any given time up to a certain level of precision. What can be predicted for a system may well depend on that level, as has been driven home by the recent work on "chaos" (or, more accurately, the sensitive dependence of a system's dynamics on its initial conditions). Because of this, and also because dynamical laws are sometimes probabilistic, prediction is often taken to be a stronger category than explanation: one can explain events without having been able to predict them.

However, things are not quite that clear-cut. Correlations without explanations (or "causation" from a more ontological or metaphysical orientation than what will be adopted in this book) often allow prediction. In genetics, this is extremely important. There is little doubt that certain statistical parameters, such as (narrow) heritability, admit prediction under certain circumstances. But, as examples given in Chapter 4 will show, high heritability, which is routinely taken as indicative of the genetic origin of traits, can occur when genes alone do not provide an explanation of the genesis of that trait. To philosophers, at least, this should come as no paradox: good correlations need not even provide a hint of what is going on. They need not point to what is sometimes called a "common cause." They need not provide any guide to what should be regarded as the best explanation.

1.5. OUTLINE

The contents of the various chapters of this book have already been alluded to in the previous sections. More systematically, Chapter 2 begins the explication of reduction and provides a short discussion of the various *formal* issues that have interested philosophers. The discussion of these issues is intentionally kept as short as possible because the thrust of this book is towards *substantive* issues. Chapter 3, which completes the general discussion of reduction, is concerned with these substantive issues. Five different concepts of reduction are distinguished on the basis of the substantive criteria that are introduced. Three of these (as discussed in detail in later chapters) are particularly relevant to genetics. Though the examples used in these two chapters are almost entirely from biology, the discussions are intentionally kept at a sufficiently general level for them to be potentially useful in other contexts in the natural sciences.

Chapters 4, 5, and 6 analyze three different types of reduction – which are different on the basis of the substantive criteria introduced in Chapter 3 – that are attempted in genetics. Chapter 4 takes up the type of reduction that is ostensibly carried out to explain phenotypic phenomena from the genotypic level with no specific assumption being made about the types of alleles or the organization of the loci that may be involved. The focus of this chapter is on the use of various concepts of "heritability" to attempt to explain phenotypic phenomena from a genotypic basis. Thus, this chapter deals with the methodologies involved in some of the most politically controversial claims of genetic reductionism, in particular, the alleged heritability of and genetic basis for IQ. Roughly speaking, this chapter is primarily concerned with the methodology of quantitative genetics.

Chapter 5 turns to ordinary population genetics and analyzes segregation and linkage analysis and other similar methods that are also used to explain phenotypic traits from a genetic basis. In these explanations, assumptions about the usual structure of the genome (such as ploidy and the linear order of loci) and the rules for the transmission of genes from one generation to another are critical. An attempt is made to show that these methods are epistemologically superior to those of Chapter 4. Nevertheless, some caution has to be used when these methods are invoked; the potential pitfalls of inferences to genetic explanation using these methods are discussed in some detail. This chapter also deals with the relation of biometry with Mendelism.

Chapter 6 takes up physical reductionism in genetics. It is argued that this is the type of reductionism that has generally characterized research in molecular biology. Several models are analyzed in detail, including one of the more spectacular successes of physical reduction in genetics, the operon model, as well as another one from molecular biology outside genetics, namely, the Monod-Wyman-Changeux model of allostery. Unlike genetic reductionism, physical reductionism has had a relatively successful career in biology. This point is emphasized because it appears to be denied by many philosophers. It is argued that much of the philosophical confusion about the viability of reduction in molecular genetics has arisen from a conflation of genetic and physical reductionism. The problems with the former have been used to argue against the success of the latter. Meanwhile, within molecular biology the reverse process has sometimes taken place. The success of the latter has been used to tout the plausibility of the former, especially by those who initiated the Human Genome Project.

Chapter 7, which masquerades as a conclusion, is relatively polemical. It notes how champions of genetic reductionism are remarkably prone to ignore many elementary biological points in their attempts to advance their intellectual (and, possibly, their political) agendas. This chapter includes a tentative attempt to define "genetic." Rather crudely, a trait is "genetic" if and only if it can be tangibly reduced to "the genes." A convention about the use of "genetic" is proposed. The purpose of this proposal is to retain "genetic" in philosophical and scientific considerations, thus rejecting the more radical proposals to drop it altogether, while fully recognizing the difficulty of defining that term. However, it is noted that dropping the term altogether will not present any great inconvenience in the pursuit of genetics, its implications, or its foundations. Here, this analysis agrees with those critics who doubt the value of "genetic." Finally, it notes some of the points that directly follow from the relatively technical analyses of the previous chapters.

Finally, it should be emphasized that while an attempt has been made here to provide a fairly comprehensive analysis of the issues that are pertinent to reductionism in genetics, no deliberate attempt has been made to review all the literature or to provide a comprehensive bibliography.[30] This book presents a particular point of view about reductionism. This is most clearly expressed in Chapter 3, where some of the distinctions that are introduced appear to be new in the literature. Throughout this book the emphasis is

14

on substantive rather than formal issues, and this book attempts to use a distinction between epistemological and ontological questions to clarify the debates about reduction and reductionism. In general, it emphasizes epistemological questions at the expense of ontological ones. Therefore, it is hardly "philosophically neutral," whatever that might mean. Whether the framework developed here is useful, or even really defensible, remains to be seen.

2

Types of reduction: Formal issues

The idea of reduction in the empirical sciences is at least as old as the mechanical philosophy of the seventeenth century, which in effect simply required that all physical phenomena be explained by, or "reduced" to, local contact interactions between impenetrable particles of matter. The properties that were attributed to these particles were size, shape, motion, and sometimes gravity. On this basis all physical properties of bodies – such as mass, weight, or their ability to reflect, refract, absorb, or polarize light – as well as all their chemical properties, were to be explained. By the middle of the nineteenth century it became clear that the mechanical philosophy – as originally conceived – could not provide an adequate basis for physical theory. However, the program of accounting for the theories, laws, and empirical facts in one scientific domain on the basis of those in another, that is, "reducing" the former to the latter, continued to play a significant role in scientific research. In fact, it was in the nineteenth century that this program began to achieve some important successes. Maxwell's theory of electromagnetism provided an implicit reduction of the laws of geometrical optics to what may be called "physical optics." Maxwell, and especially Boltzmann, attempted to reduce the thermodynamic laws to the kinetic theory of gases. Meanwhile, Helmholtz and his associates developed an experimental program to found biology upon physical principles. These programs achieved at least partial success.[1]

The logical empiricist program in the early part of this century at least implicitly recognized the importance of reduction in the development of science. This was one possible route (though not the one that they generally endorsed) to the "unity of science" that they cherished.[2] Systematic analysis of reduction only began with Nagel (1949), who regarded it as a species of intertheoretic explanation.[3] A reduced theory is explained by a reducing theory that is presumed to be more fundamental. A similar analysis was independently developed, almost simultaneously, by Woodger (1952). In the late 1950s this analysis was rejected and alternatives offered by Kemeny and Oppenheim (1956) and Suppes (1957); these will be discussed in § 2.3. Suppes's approach has since been extended and advocated by Sneed (1971), Stegmüller (1976), Balzer and Sneed (1977), and, explicitly in the context of genetics, by Balzer and Dawe (1986a, b). These will be considered in some detail below. Meanwhile, the unity of science through reduction was explicitly proposed, though only as a "working hypothesis," by Oppenheim and Putnam (1958).

Nevertheless, it is safe to say that almost all accounts of reduction that view it as a relation between theories are variants of the Nagel-Woodger approach that has provided the framework in which disputes about reduction have taken place. Most of these disputes have been about *formal* issues, the "logical" form of reduction. Even the dispute between those who require that reduction be a relation between theories and those who suggest that it is not has largely been about formal issues. These formal issues are the subject of this chapter but, before their analysis begins, two distinctions central to that analysis, between *formal* and *substantive* issues, and between *epistemological* and *ontological* questions, will have to be clarified. After this chapter the formal issues will be relegated to the background and the rest of this book will concern itself with substantive concerns about reduction.

Thus, the point of departure of this book will be the analyses of reduction that are similar to those of Nagel and Woodger. Even though these will largely be ignored after this chapter (though not as much for being incorrect as for being unilluminating), it might still appear a startling starting point given that the general logical empiricist account of science has long been rejected by most philosophers of science. Even if the "scientific realism" of the 1960s and 1970s has so far not lived up to its initial promise,[4] since the early 1980s many philosophers of science including van Fraassen (1980) and Cartwright (1983, 1989) have moved to views of science based less on theories than those of the logical empiricists. This book is consonant with those views of science that are generally skeptical about the status of theories. Moreover, as will be apparent below, by being concerned with detailed mechanisms rather than broad generalizations, and because of worries about the ontological issues raised by approximations and many similar points, much of what is said here is particularly consonant with the view of science advocated by Cartwright (1989).[5] Nevertheless, these new views have largely ignored the issue of reduction. Consequently, as in the case of others who have worried about reduction in biology in the recent past, the point of departure, as so often remains the case in the philosophy of science, remains the work of the logical empiricists.[6]

Genetics and reductionism

2.1. FORMAL AND SUBSTANTIVE ISSUES

The strategy followed in this book will be to make a distinction between formal and substantive issues in the philosophy of science, discuss the formal issues with respect to reduction perfunctorily, thus implicitly suggesting that the substantive issues are more interesting and important, and then move on to them.[7] Part of the thesis will be that past discussions of reduction have largely concentrated on formal issues and, therefore, have usually ignored substantive issues, thus missing much of what is subtle about reduction.[8] The last claim should not be overinterpreted as suggesting that substantive issues have altogether been ignored in the the past work on reduction. Nagel (1961), in particular, discussed substantive issues with much insight. In fact, the distinction being made here is almost identical to that he made between "formal" and "nonformal" conditions for reduction. It is different only insofar as both sets, of formal and substantive issues, are much broader than what he could envision in 1961 before the logical empiricist consensus on the structure and methodology of science fell apart.[9]

Without the wisdom of hindsight, before the logical empiricists, it would be hard to delineate clearly a set of formal issues in the philosophy of science. Even Pearson, one of the founders of biometry and perhaps the most prominent positivist of his generation, spends almost the entirety of *The Grammar of Science* (1900) discussing in detail the contents and meaning of scientific theories and procedure, paying almost no attention to their representation or other questions of form. Pearson's positivism consisted almost entirely of a rejection of causality in favor of correlations. The logical empiricists (the new positivists) changed all that. The primary weapon that they deployed in their battle with traditional philosophy was the apparatus of mathematical logic constructed by Frege and Russell. In the late 1920s, especially in the context of German philosophy, this led to a substantially higher standard of rigor and clarity.

Then, at some point in the 1930s, probably misled by their faith in the ability of mathematical logic to clarify all scientific and philosophical issues, the logical empiricists decided that the philosophy of science (which, for most of them, had become all that there was to philosophy) was to concern itself solely with an analysis of the language of science, that is, the form of scientific sentences. That move exerted a pernicious influence on the field for a full generation (including the period when explicit discussions of reduction began). Worries about linguistic form dominated the analytic philosophy of science at the expense of more interesting questions of what (to invoke a relatively neutral term) might be called "natural philosophy,"

18

questions about the implications of, and the assumptions behind, scientific theories, models, explanations, etc.[10]

The single figure mostly responsible for this philosophical move was Carnap who, in *The Logical Syntax of Language* (1937) and other writings from the 1930s, explicitly urged that philosophy become or, rather, be replaced by the analysis of the syntax of the language of science. After Tarski convinced him of the viability of semantics, Carnap modified this proposal in favor of the analysis of the syntax and the semantics of the language of science (Carnap 1942). To be fair to Carnap, his conception of semantics was remarkably broad. After Tarski's intervention, truth was part semantics. Meaning, which was part of syntax in *The Logical Syntax of Language*, joined truth in semantics. In the 1950s, after Carnap had come to be convinced of the possibility of an inductive logic, even confirmation became part of semantics. It is easy enough to imagine that any genuinely scientifically relevant issue would have been accepted by Carnap as a part of semantics.[11]

However, in practice, the semantic concerns of logical empiricism's descendants – the analytic philosophers of science – have often been narrowly restricted to those semantic questions that can also be regarded as questions of form.[12] Syntactic issues and those semantic issues that have been framed as questions of form will be called "formal" here. These include the logical form of scientific laws and theories, the structure of scientific explanations and inference, the form of counterfactuals, and so on. Other (scientifically or philosophically) relevant issues, which will generally concern the interpretation of scientific arguments, especially what they imply and what they assume about the world, will be called "substantive." The following three points should help clarify this distinction.

(i) It is not the distinction between syntax and semantics, at least as that distinction is conventionally framed. As has been explicitly stated, some semantic issues are formal. For instance, the question whether some statements are identities is a formal question (see § 2.5).

(ii) Many of the substantive issues that will be discussed in this book would be admitted as "semantic" (at least) by Carnap, though some (such as the value of carrying out a reduction or the desirability of ontological elimination) would no doubt be relegated to the "pragmatic" netherland, that is, to the realm of questions that, however important for science, were neither syntactic nor semantic and, therefore, outside the purview of philosophical analysis.[13]

(iii) In at least one sense, this distinction cannot be regarded as absolute. Nothing, in principle, prevents the incorporation of some substantive concern into the formalism of an argument. For instance, one substantive issue that will receive detailed treatment in the next chapter is that of whether a system can be represented, for the purpose of an explanation, as a composite of spatially contained distinguishable parts. It should be easy enough to formalize any such assumption about representations and to incorporate that formal characterization into an argument. Thus, a substantive issue would be transformed into a formal one. Therefore, the distinction being made here is relative to a particular formalism (or set of formalisms). The context will indicate what that (background) formalism is. In general, it will be the logical empiricists' explication of theories, and analysis of intertheoretic explanation.

At the height of logical empiricism, formal issues dominated discussion, not only with respect to general questions in the philosophy of science, but even with respect to questions arising from the special sciences. In biology, this was primarily reflected in the work of Woodger (e.g., 1937, 1952) who was one of the very few from within the logical empiricist camp who paid any attention to biology. Since the 1960s, at least with respect to issues arising from the special sciences, philosophical discussion increasingly began to incorporate substantive issues, often adopting the analytic techniques native to a given science rather than casting everything in the form of mathematical logic. Discussions of reduction in biology were a striking exception to this rule perhaps because the issue of reduction seems to arise in the same manner in different sciences and, therefore, it appeared that it should be analyzed in a way that is independent of any special science. It does not follow from this, that formal issues should dominate the discussion. However, historically, that is what happened, and this chapter will analyze these formal discussions to glean whatever insight that they provide. However, before that analysis can proceed, a second discussion, between ontological and epistemological questions, will be important.

2.2.   ONTOLOGICAL AND EPISTEMOLOGICAL QUESTIONS

Shimony (1987) explicitly makes a distinction that, though implicit in many earlier discussions of reduction, has often enough been ignored.[14] This distinction is between ontological and epistemological questions about reduction. In this discussion it will be assumed that it is sensible to talk about ontology in a fashion that is at least independent of the precise assumptions being made about the systems and laws at both the levels of the explanans and the explanandum.[15] As Shimony puts it, in the context of a discussion of the relations between "parts and wholes," which potentially provides one of the most interesting of the types of reduction that will be considered in Chapter 3 (especially in § 3.4), the problem

> has an ontological aspect, which concerns the properties of the components and the composite system without explicit consideration of how knowledge of them is obtained. Among the ontological questions are the following: . . . If the properties of the components are fully specified, together with the laws governing their interactions, are the properties of the composite system then fully determined? In particular, are those properties of composite systems which are radically different from those of the components, and which might properly be characterized as "emergent," also definable in terms of the latter? . . . [It] also has an epistemological aspect. Suppose that the most precise and best-confirmed laws turn out to govern relatively simple systems . . . but the systems of interest are enormously complicated combinations of simple components. Then there will be insuperable experimental difficulties in gathering knowledge about all the initial conditions of the parts, and insuperable mathematical difficulties in deducing from the basic laws the properties of the composite system. To what extent can the composite system be said to be understood in terms of the laws governing its parts? And if there is independent phenomenological knowledge of laws on a coarse level of description, how do we know that these are in principle derivable from the laws on a finer level? (1987, pp. 399–400)

There is, obviously, no reason why this distinction should be confined to the particular type of reduction ("parts and wholes") for which it is being made. All that is required in these questions is to replace the discussion of a composite system by that of a system in one realm (that of the target of an intended reduction, i.e., of the explanandum) and the discussion of component systems by that of systems in the other realm (from which reduction would take place, the realm of the explanans).

The distinction between explanation and determination is an important instance of the (more general) distinction between epistemological and ontological issues. From Shimony's point of view, and from the point of view adopted here, explanation is obviously an epistemological issue.[16] However, the question whether a set of entities and interactions at one level determines those at another is an ontological question. Even if some such determination holds, it does not ensure that an explanation is available or even forthcoming. (This, in fact, is close to the position of those who suggest "supervenience" rather than reduction as the appropriate relation between domains in many contexts; this will be discussed in § 2.6.) That these issues can and in practice often are kept distinct is a point that will be emphasized later in this section.

Simple though this distinction may be, it helps avoid some apparent paradoxes. It shows how it is possible to deny the reduction of biology to physics, as many have done, from Bohr (1933) and Delbrück (1949) to Mayr (1982), without endorsing any form of vitalism.[17] The denial, in all of these cases, is a denial of reduction understood as explanation (as it is construed in this book); it is a negative answer to an epistemological question. A denial of this sort could obviously be – and often is – motivated by formal reasons. For instance, it could be argued that the putative explanations in some field do not satisfy the strictures of an assumed formal explication of reductionist explanation. In the case of genetics, exactly this claim was made by Hull (1972, 1974) who argued, among other things, that the explanations offered by molecular biology did not satisfy the formal requirements of Schaffner's (1967b) model of reduction.[18]

However, what is much more interesting is that a denial of this sort could also be motivated by substantive reasons (and these were the type of reasons that motivated Bohr, Delbrück, and Mayr). Because of the complexity that Shimony refers to, or possibly because of an *in principle* difficulty in obtaining sufficiently precise initial data, explanations may fail for substantive reasons with no implication about the ontological issue of whether "something else" is going on. In contrast, vitalism, as conventionally understood, is an ontological doctrine: it claims that there are processes occurring in living matter that do not occur in nonliving matter. Thus, even denying the possibility of physical explanation of biological phenomena in general does not amount to an endorsement of vitalism.

It is, therefore, easy to be an "ontological reductionist" without being an "epistemological reductionist." Is the converse possible; that is, is it possible to endorse epistemological reduction but not ontological reduction? At first sight, if one is at all willing to talk about ontology in this broad fashion (which this discussion assumes), this might appear impossible. Surely, one has to assume that the entities and properties occurring in the explanans exist, and that they give rise to those in the explanandum? The issue is not quite this simple. Explanations routinely involve approximations and idealizations, some of which, in turn, involve counterfactual assumptions (see Chapter 3, § 3.4). If a very good explanation (affording very accurate predictions) of the interaction of a DNA segment and a protein requires the treatment of atoms as rigid balls, does the acceptance of this explanation require a real commitment to the position that there are atoms of this sort, and this is what the system is made up of? Presumably not, because atomic physics (that is, the quantum mechanics of atoms) denies the possibility that such atoms exist. Assuming rigid-ball atoms in the explanation is, at the very least, a counterfactual approximation.

Many other examples of this sort will be discussed in the chapters that follow (e.g., Chapter 3, § 3.2 and Chapter 6, § 6.3). In such examples, if ontology is to be taken seriously, the connections between what is assumed to exist in the reducing realm (that of the explanans) and in the reduced realm (that of the explanandum) are far from clear. It is often little more than a matter of faith that a consistent story, which relies entirely on assumptions from the reducing realm, can be told. One could choose to say in such a situation that it is the explanation that has failed whenever any potentially counterfactual move is made, even when, for example, that move is as innocuous as assuming that rigid bodies can exist, or letting some parameter, such as the number of (genetic) loci, go to ∞. However, the denial of explanatory success in such cases seems far more counterintuitive than avoiding ontological reduction altogether, which is the option that will be adopted in this book.[19]

Distinguishing between ontological and epistemological questions does not, of course, deny that there might well be deep connections between the ontological and epistemological assumptions that may be made during the course of a scientific investigation. The entities (if any) to which one is committed might well influence what type of explanation one seeks. To make this claim a little stronger, there is a sense in which every ontological assumption comes with an associated epistemological program, that of using the assumption to tackle potential explananda. Conversely, the explanatory success of a theory or research program might well increase one's confidence in its ontology, that is, the entities that are referred to in its claims and assumptions. However, these are not necessary connections. One might well prefer an ontology that accepts only the most fundamental objects of physics as "real" and yet pursue upper-level explanations (as, for instance, the biochemical explanations of biological phenomena) for pragmatic reasons such as the ease of computation or representation. Explanation, after all, has a pragmatic component; otherwise one would never know when to stop explaining. Conversely, one could pursue different modes of explanation and have no concern about ontology.

This book will fall in the last-mentioned pattern. After Chapter 3, there will be virtually no overt discussion of ontology. The decision to ignore ontology is based partly on philosophical reasons: the problems raised by approximations and a doubt whether any genuinely sensible ontological questions can be asked in answerable form. However, the main reason for ignoring these questions here is simply that, in the genetic context, they are not interesting.[20] No one believes in vitalism any more; and even if a trait is fully explained from a genetic basis, no one is likely to suggest that all considerations about the trait be simply replaced by those about the genes involved.

2.3. THEORIES VERSUS MECHANISMS

An important dispute about reduction has been about the question whether reduction must be considered a relation between theories.[21] An answer to this question depends on what is taken to be the proper criteria for identifying a scientific theory. It will be treated here as a formal rather than substantive question only because that is the most convenient way to organize and analyze those philosophical disputes about reduction that have focused on the role of theories. This is a consequence of the fact that these have largely been general disputes about the form of explanation. However, a plausible case can be made for the point of view that the question of what constitutes a scientific theory is a substantive question.

Certainly, it is clear that what constitutes a theory cannot be regarded as a syntactic question. Those philosophers who have followed the logical empiricists into the analytic tradition in philosophy generally conceive of theories as consisting of laws, each of which is represented by a universally quantified (at least approximately) true sentence of the form: $(\forall x)(Fx \rightarrow Gx)$, where x ranges over some appropriate universe of discourse and F and G are predicates. However, this form does not distinguish between the so-called accidental generalizations and what may be called the laws of nature. To do so requires extrasyntactic criteria. What these criteria should be has never been fully settled.[22] At the very least, a "law of nature" should help sysematize a body of knowledge; have some relation with other laws, thus forming a theory; and do so in such a way that the theory provides a succinct description of the much larger body of knowledge (that is being systematized).[23] There is no reason to suppose that these criteria can be fully formalized in any "natural" way; at least, there has been no real success toward that end.[24] Using the distinction between formal and substantive issues that was made above, this would imply that the question of what constitutes a scientific theory becomes a substantive question (though one on which very little progress has been made).

However, the dispute over the role of theories in reduction has not been about this point, though a belief that there is no theory (in this sense) in classical genetics [for example, by Kitcher (1984)] or molecular genetics [for example, by Hull (1972)] has sometimes been used to reject the claim that the former had been, or was being, reduced to the latter. Purely critical moves of this sort provide no alternative analysis of reduction.[25] Rather, they constitute an acceptance of the standard (Nagel-Woodger) view of reduction, and the disagreement is about the status of a particular case. More importantly, in this context, both Hull and Kitcher not only reject the claim that scientific laws exist in some context (a substantive claim) but also reject the claim that connections between the reduced and reducing theories can be put into the appropriate form (see the discussion in § 2.5), which is a formal question.

In contrast, Wimsatt (1976b) provides an entirely different account of reduction – though only for the reduction of wholes to parts, which he calls "compositional" reduction – which denies a necessary role for theories and, in effect, presents an alternative analysis of explanation. It will be taken up in the next section. The rest of this section will describe the Nagel-Woodger model, its important modification by Schaffner, and, for the sake of completeness, two other models of reduction (one due to Kemeny and Oppenheim and the other due to Balzer and various collaborators) that also view it as a relation between theories. The discussion of the Nagel-Woodger and Schaffner models will be somewhat detailed, not only to be able to delineate, clearly, the extent of the difference between them and Wimsatt's model, but also because most of the other formal disputes about reduction have occurred in the context of these models.[26]

Nagel (1961) assumes that reduction is a relation between two theories, the reduced theory (T_1) and a reducing theory (T_2), the relation being one of explanation, where explanation is to be understood as the logical deduction of T_1 from T_2.[27] Both T_1 and T_2 are assumed to be formalized in first-order logic. Nagel called this requirement (of deduction) the "condition of derivability" (p. 354). If T_1 contains no term not already in T_2, such a deduction can be immediately attempted. According to Nagel, such deductions are *homogeneous*. If, however, T_1 does contain such terms, then the terms in T_1 must somehow be connected to those in T_2. Such reductions are *heterogeneous*, according to Nagel (p. 342).[28] The requirement that the terms be connected, which is a prerequisite for any attempt at deduction, is Nagel's "condition of connectibility" (p. 354). The nature of the connection between the terms of T_1 and T_2, usually called "bridge laws," "reduction functions," or "connectibility assumptions," has been a continual source of disagreement and will be treated in some detail in § 2.5. Suffice it here to note that, for Nagel, their syntactic form could be that of a universally quantified biconditional or conditional sentence, and they were to be interpreted as conventions or factual statements depending on the context. Thus, reduction is for Nagel a purely epistemological issue with no necessary ontological commitment.[29] Woodger (1952) arrived at a similar model independently – and, indeed, when Nagel's analysis was still somewhat rudimentary. Woodger's model is actually a more restricted version of Nagel's for two reasons: (i) Woodger restricts attention to the case when the vocabulary of T_2 is a proper subset of the vocabulary of T_1, that is, to a rather special case of homogeneous reduction; and (ii) he requires the bridge laws to be biconditionals.

This model of reduction, which came to be regarded as the "Nagel-Woodger" model,[30] has three appealing features, at least to logical empiricists and their followers: (i) in logical form, it is quite simple; (ii) the model of explanation that is invoked is only a slight modification of the standard deductive-nomological model of explanation that, especially during the heyday of logical empiricism, was generally accepted (at least as far as formal questions are concerned)[31]; and (iii) Nagel's emphasis on epistemological over ontological issues was consistent with the antimetaphysical attitude of the logical empiricists. Nevertheless, even at the formal level, there are many obvious facts about actual scientific reductions that this model cannot incorporate. Perhaps the two most blatant of these are that (i) not only are scientific derivations (such as those that usually form part of any explanation) almost never logical deductions, they cannot be reconstructed as such, even when they are fundamentally mathematical – the derivations often involve so-called "physical approximations" and other heuristics[32]; and (ii) even after all this heuristic reasoning, what is obtained at the end is not usually the reduced theory, but something that – with luck – is recognizably similar to it.

The former feature of reductions has never received the kind of attention that it deserves. An attempt to treat it will be made in the next chapter (Chapter 3, § 3.4); however, because these heuristics and approximations are never fully formalized, it will be treated as a substantive issue. The latter feature led Feyerabend (1962) to declare a general incompatibility between putative reduced and reducing theories and to argue against the possibility of reductions.[33] This feature, and also the point that an attempt to carry out a reduction might result in a change in the reducing theory – after all, the potentially reduced theory (as explanandum) is part of the evidence that the reducing theory must account for – led Schaffner to modify the Nagel-Woodger model, but retain its spirit.[34]

The basic modification is the following. Instead of T_1 and T_2, one has two new theories T_1' and T_2', and Nagel's conditions of connectibility and derivability are now to be satisfied between T_1' and T_2'.[35] T_1' is supposed to be "strongly analogous" to T_1, and to correct the latter, thereby answering Feyerabend's objection: what the reducing theory does is correct the reduced theory, and this is what one should expect since the former (which is doing the explaining) is epistemologically more fundamental than the latter. T_2' is a successor theory to T_2. It incorporates the changes induced in T_2 by the attempted reduction (and, possibly, other sources). What emerges is a much more dynamic view of scientific theories than what Nagel or Woodger had presented. Schaffner continued to think of the various theories as axiomatized in first-order logic. Unlike Nagel, however, he insisted that not only must the bridge laws have the form of biconditionals, but that they also be synthetic identities. In requiring that they be synthetic, he removes one of Nagel's possibilities, that they be conventions, and then conflates "factual" with "synthetic." In requiring them to be identities, he goes beyond purely epistemological concerns and issues an ontological criterion that would probably have been unpalatable to Nagel.

These bridge laws will be discussed in detail later ($\S$ 2.5). The other controversial element in Schaffner's model is that there seemed to be no context-independent characterization of the relation between T_1 and T'_1 or that between T_2 and T'_2, let alone a formal one.[36] In such a circumstance, no model of reduction that relies only on formal criteria for delineating theories and relations between them can be constructed. Moreover, as Wimsatt (1976b) has pointed out, both of these relations are intransitive similarity relations. This leads to the possibility that when, finally, the successful deductions are carried out, the T_2 that is involved may no longer be interesting as a potential reducing theory different from T_1. However, by liberalizing the Nagel-Woodger model to allow for some dynamism in the view of theories, and by answering Feyerabend, Schaffner's model has obvious advantages over the former, while retaining much of the logical empiricist orientation toward science, and whatever insight that it may contain.

In passing at least, two other types of reduction models that also view reduction as a relation between theories should be noted. However, for the reasons that will be given below, they will not be considered any further in this book. Both of these view the relation between the reducing and the reduced theories as indirect. The first, due to Kemeny and Oppenheim (1956), is actually a model of theory replacement, rather than reduction (in any intuitive sense), though they claim it is the latter. As they put it:

> Scientific progress may be divided into two types: (1) an increase in factual knowledge, by the addition to the total amount of scientific observations; (2) an improvement in the body of theories. . . . An especially important case of the second type is the replacement of an accepted theory . . . by a new theory . . . which is in some sense superior to it. Reduction is an improvement in that sense. (pp. 6–7)

The new theory must be better systematized which, from their point of view, means that it is simpler or has greater (explanatory) strength. There is, thus, no direct relation between the two theories. However, since both theories must be (logically) related to the terms and predicates of the same set of observation sentences, there is a weak and indirect relation (p. 16).

As Schaffner (1967b) showed, the Kemeny-Oppenheim model puts much weaker restrictions on reduction than either the Nagel-Woodger or Schaffner model. Moreover, by foregoing any necessary relation between the theories, it does not allow a reduction between them (as has already been mentioned): the new theory does not, in any ordinary sense, explain the old. At best, it is a model of theory replacement, but with precious little detail about what systematization is to consist of, how simplicity is to be calibrated, or even compared to explanatory strength, and so on. It will not be considered any further here.

The third – and relatively novel – type of model of reduction, developed most systematically in the genetic context by Balzer and Dawe (1986a, b), is based on a scheme originally due to Suppes (1957).[37] Following Suppes (1957), as well as Sneed (1971) and Stegmüller (1976), Balzer and Dawe adopt a method of reconstructing theories using (informal) set-theoretic predicates (rather than in first-order logic). What is gained, thereby, is a significant simplicity in formalization. Once a theory is thus axiomatized in set theory, its intended applications are specified, and then the empirical content of the theory becomes the claim that all these intended applications are models of the theory.[38]

The basic intuition about reduction in this approach is that one theory reduces to another if, given models of the two theories, there exists a relation between such models (the "reduction relation") that ensures that every model of the reduced theory is also a model of the reducing theory. The existence of this formal relation, however, does not ensure that the two theories are not about entirely disparate subjects. Balzer and Dawe (1986b), therefore, construe it as a necessary but not sufficient condition for reduction. To ensure sufficiency, they impose two further conditions: (i) as a nonformal condition, they simply require that the two theories be about the same object; and (ii) they require that the reduced theory be deducible from the reducing one.

However, they do not construe deduction syntactically (which, at the formal level, would have left them presenting little more than a variant of the Nagel-Woodger model). Rather, they put two new conditions on the models and show that these are satisfied for the cases of interest. The first is the existence of the reduction relation (noted before).[39] The second is that there exists, for all models of the reducing theory, at least one model of the reduced theory that has the reduction relation to it.[40] However, these two conditions are sufficient to ensure deducibility (even in their sense) only if the theories in question can be formalized in first-order logic or some formal language that is only slightly more general.[41] Even if such a formalization is possible – and having to test it destroys the attractive simplicity of the set-theoretic approach – it does not logically guarantee (ordinary syntactic) deducibility. Therefore, in this model there is no guarantee that an explanation takes place when a reduction is obtained.[42]

This is sufficient epistemological reason not to pursue the Balzer-Dawe model any further in this book. However, it should also be emphasized that this model does not provide any new insight about ontological issues either. The requirement that the various theories be about the same subject is introduced by fiat (as a nonformal requirement). The isomorphism of the various models does not guarantee any identity of interactions (or processes). One is left wondering what all this formalism was developed in aid of. Suffice it, finally, to note that this formalism does not even capture actual instances of scientific reasoning to the extent that, for instance, the deductive-nomological model of explanation does.[43]

2.4. THE FORM OF EXPLANATION

The most systematically developed model of reduction that is significantly different from the Nagel-Woodger and Schaffner models, that is, that cannot be regarded as a variant, is that of Wimsatt (1976b). This is a formal model – a general model of reduction based on Salmon's (1971) account of explanation.[44] It rejects the basic Nagel-Woodger or Schaffner model of explanation, in particular, that explanation – and, ipso facto, reduction – must have a theory (or a law that forms part of a theory) as the explanans. Reduction is no longer a relation between theories. Two other models of reduction, those of Kauffman (1972) and Sarkar (1989, 1992), also do not view explanation and, similarly, reduction as necessarily a relation between, or even involving, theories. Thus, these models, too, reject the formalism of the Nagel-Woodger model. However, Kauffman (implicitly) and Sarkar (explicitly) do not tie their models to any particular formal analysis of explanation; Kauffman does not even use the term "reduction," preferring "articulation of parts explanation" (p. 257). Sarkar – using the strategy that is being followed in this book – only attempts to elaborate what additional criteria an explanation must satisfy to be a reduction. Both Kauffman and Sarkar are concerned with substantive issues, and the possible insights to be gleaned from their models will be incorporated into the next chapter. Attention, here, will be confined to Wimsatt's model, which is developed formally.

Wimsatt (1976b) distinguishes between reductions that are explanations with "compositional redescription" and those without, that is, those that involve "inter-level" explanation and those that remain "intra-level."[45] His new model of reduction is only supposed to the former case (that is, interlevel with compositional redescription). He begins by modifying an account of explanation due to Salmon (1971) that confines itself to the explanation of particular cases, and rejects the notion that explanations should be regarded as arguments. Salmon's analysis is geared primarily toward the understanding of statistical explanation of individual events: scientific explanation is the search for statistically relevant factors, and an explanation is best seen as a list of such factors, with associated conditional probabilities for the occurrence of the event that is to be explained. Deterministic explanations, including those that are putatively captured by the deductive–nomological model, are degenerate cases on this account, when the probability of one of the events becomes 1, and all others 0.

Salmon (1971) views a scientific explanation as an attempt to answer a question of the form: "Why does this x which is a member of A have the property B?" (p. 76). The answer, which explains why x has occurred, begins by partitioning the reference class A into subclasses that are "homogeneous with respect to B," that is, there is no way (even in principle) to effect a statistically relevant partition of any of them with respect to B without knowing which elements of the subclass have B. Next, probabilities of having B in each of these subclasses must be specified, and finally the particular partition that contains x must be indicated. As Salmon puts it:

> More formally, an explanation of the fact that x, a member of A, is a member of B would go as follows:
>
> $$P(A.C_1, B) = p_1$$
> $$P(A.C_2, B) = p_2$$
> $$.$$
> $$.$$
> $$P(A.C_n, B) = p_n$$
>
> where
>
> > $A.C_1, A.C_2, \ldots, A.C_n$ is a homogeneous partition of A
> > with respect to B,
> >
> > $p_i = p_j$ only if $i = j$, and
> >
> > $x \in A.C_k$[46] (1971, pp. 76–77).

Consider any property C that is used to partition A. Some other property D "screens off C from B in reference class A iff (if and only if) $P(A.C.D, B) = P(A.D, B) \neq P(A.C, B)$" (p. 55). Using this definition, Salmon introduces the "screening-off rule:"

> When one property in terms of which a statistically relevant partition in a reference class can be effected screens off another property in terms of which another statistically relevant partition of that same reference class can be

effected, the screened-off property must give way to the property which screens it off. (p. 55)

The former then becomes "irrelevant and no longer has explanatory value" because, in a sense, the new property provides a deeper explanation of the event.

Salmon intended his model to provide a general account of explanation. Any direct attempt to apply it to reduction leads to five transparent difficulties, the first two of which also constitute a general problem with it as a model of explanation (Sarkar 1989): (i) these considerations do not, in any direct way, allow the explanation or reduction of laws[47]; (ii) the account of deterministic explanation is counterintuitive[48]; (iii) if C and D are provided by the reduced and reducing theories, generally (and it does not matter whether these are full-fledged theories), there is no sense in which D explains C; (iv) such an analysis is necessarily eliminative simply because C becomes irrelevant and disappears; and (v) there is no distinction in this analysis between reduction or explanation and replacement (which has figured prominently in disputes about particular cases of scientific change, including the transition from classical to molecular genetics).

Wimsatt (1976b) introduces two modifications to this analysis in an effort to capture those reductions in which some property of a composite system is being explained by the properties of its constituent parts:[49] (i) by fiat – that is, with no argument or further analysis – he replaces Salmon's "statistical relevance" with "causal relevance"[50]; and (ii) when D screens off C from B in A, Wimsatt also introduces a notion of "effective screening off" in the reverse direction:

C effectively screens off D from B in reference class A if (and perhaps not only if):
(a) $P(A.C.D, B) = P(A.D, B)$
(b) $P(A.C.D, B) \cong P(A.C, B)$ or $P(A.C.D, B) \neq P(A.C, B)$
(c) cost $(D) \gg$ cost (C) {D is enormously more expensive
                    information to get than C.}
(c′) D [consists of] a *compositional redescription* of C[51] (1976b, p. 702).

Reduction occurs, according to this scheme, if and only if C effectively screens off D. Note that the definition of "C effectively screens off D" includes the requirement that "D screens off C" because of clauses (a) and (b), interpreting both "$\cong$" and "$\neq$" as "$\neq$" (in Salmon's usage).

Wimsatt intended (c) and (c′) as alternatives. Clause (c) introduces the concept of a cost of explanation. Wimsatt does not provide any definite method for its determination in practice, but the basic intuition is that obtaining explanations using the "reducing" list D is much more complicated than using the "reduced" list C. Sarkar (1989) suggested that both clauses (c) and (c$'$) should be used in such a definition of reduction, with (c) being a "Condition of Utility" and (c$'$) being a "Condition of Reduction." The former name is suggested by the fact that (c) would provide the rationale for the continued use of C, thereby making its elimination and replacement by D implausible on pragmatic grounds.[52] The latter name is suggested by the fact that only (c$'$) ensures that the explanation being offered by D is reductionist in the sense that it invokes the constituent parts of the composite system to which C refers.

By the introduction of "effective screening-off" Wimsatt removes the last two of the five difficulties with Salmon's account that were noted above: there is now no question of elimination or simple replacement. However, the first three difficulties – that there is no straightforward way to extend this definition to the reduction of general claims (rather than individual instances), that the account of deterministic explanations is counterintuitive, and that there is no sense in which reducing theories explain the reduced ones – remain. The last of these difficulties can be partly removed by a rather simple modification: require that D provide a partition of all the classes $A.C_i$. At least, in that case, D consistently provides a "finer graining" of the classes than C; this point will not be pursued here.

All of these difficulties can be avoided by disclaiming any special role for laws, theories and deterministic explanations in reductions involving explanations of wholes from parts. However, any such solution cannot be satisfactory simply because, not only are laws and theories often involved in such reductions, at least at one level, but also many such explanations, including many of those in molecular biology, are sometimes deterministic. The difficulties with Wimsatt's account that have been noted arise, basically, from the form of Salmon's model of explanation. However, as has been emphasized before, the questions of when a body of scientific knowledge is to be regarded as consisting of laws and theories, how these are to be formalized and represented (in philosophical discourse), and how any type explanation should be formally explicated can be separated from the more specific question of when an explanation is also a reduction.

2.5. THE FORM OF THE CONNECTIONS

Probably no aspect of reduction has been as controversial as the nature of the connections between the reduced and reducing theories (or descriptions). This will be treated here as a formal question. The radically different forms of the Nagel-Woodger or Schaffner model and the Wimsatt model would seem to require that the question be treated differently in these two cases. However, the issues that have been debated impinge upon them in the same way, and Wimsatt uses the same concepts and terms to discuss these connections as Schaffner. First, and most easily disposed of, is the question whether the connections are conventional or factual, which are the two possibilities that Nagel (e.g., 1961) left open. The difference between these two possibilities is not to be taken to be unbridgeable: Nagel argues that the context in which the connections were introduced would determine which of them is more appropriate. For instance, according to him, if the connection being introduced is the first link between the reducing and the reduced theories, it will at this stage have the status of a convention. In other circumstances the connections will be factual hypotheses demanding empirical support before being accepted.

Nagel's attitude was consistent with the liberalized – and relatively holistic – empiricism of the 1950s, which demanded only that entire theories, along with their "correspondence principles," receive empirical support from observation sentences.[53] The analogy, here, is between the correspondence principles and the connections (in the case of reductions), with the reduced theory playing the role of the observation sentences.[54] Schaffner (1967b), however, shows less tolerance and demands that these connections be "synthetic." In practice, he conflates "synthetic" with "factual." This choice of terminology is unfortunate, since it appears to make the difficulty of drawing a "fact–convention" distinction for individual sentences independent of context (which Carnap, Nagel, and the logical empiricists usually acknowledge) into that of drawing the "analytic–synthetic" distinction, about which there was, and is, far less consensus.

The most compelling reason for regarding these connections as factual was given by Wimsatt (1976b), who pointed out that the search for these connections plays a crucial role in scientific research.[55] As an example, he discussed Sutton's (1903) and Boveri's (1902) identification of Mendel's factors as chromosomes which, besides having been "discovered" empirically, is also not strictly true in the light of later developments. At the very least, these discussions make it clear that these connections are not only routinely factual, they become the locus of empirical investigation.

Whether these connections should also be regarded as "identities," as both Schaffner and Wimsatt require, is an entirely different matter.[56] Leave aside the ontological connotations of "identity" for later consideration. In form, an identity statement is a biconditional. Thus, in insisting on identities, Schaffner again opts for less tolerance than Nagel, who allowed both biconditionals and conditionals. Neither Schaffner nor Wimsatt offers any logical ground for the superiority of the biconditional; Wimsatt at least offers some justification by arguing that the search for such identities plays an important heuristic role in scientific research. This heuristic value arises because identity is a transitive relation, and thus a series of identity hypotheses across several different levels of organization permit complete identification of entities at different levels that may be far removed from each other. This, for Wimsatt, provides a method for testing laws or regularities at various levels; "[i]dentifications are [thus] an effective guide for theory elaboration" (p. 700).

This argument is not particularly convincing, and it involves what is ultimately an equivocation about "identity," between the use of that term in its strict sense or only as referring to "identification." The conditional is also transitive [that is, $(((\phi \Rightarrow \psi) \land (\psi \Rightarrow \chi))) \Rightarrow (\phi \Rightarrow \chi)$ is a logical truth] and at least partial identification can be achieved through conditional statements. For instance, a sentence of the basic form

(i) "if ζ is an allele, then ζ is located on a chromosome"

arguably offers as much identification of a gene as is necessary for genetical investigations and is certainly sufficient for the kind of reduction that Wimsatt discusses. As the discussion in Chapter 5, § 5.3 will show, segregation analysis, one of the standard techniques for reduction in classical genetics, does not make any identity assumption that is stronger than this conditional statement.

Now suppose that the theory or description at the level of chromosomal mechanics allows parts of chromosomes to be exchanged with a certain probability during meiosis (crossing-over), and that this probability can be measured (or inferred), at least at a gross level, for definite parts of chromosomes. This could be formalized as

(ii) "if ξ is a part chromosome Γ and ν is a part of (its homologous) chromosome Λ, then after meiosis, with probability r, ξ will be a part of chromosome Λ and ν will be a part of chromosome Γ."

Now, (i) and (ii) do not allow much to be inferred. What is required are sentences of the form

(i′) "if ζ is the allele A, then ζ is located at the part ξ of chromosome Γ, and if μ is the allele B, then μ is located at the part ν of the chromosome L."

In any reasonable scheme of explanation, the recombination of alleles ζ and μ, that is, a statement of the form

(iii) "the alleles ζ and μ have a probability, r, of changing chromosomes during meiosis,"

must be capable of being explained from (ii) and (i′), with the latter providing the connection between "allele" and "part of chromosome."[57] Once (iii) is obtained, as will also be apparent from the discussion in Chapter 5, linkage analysis, another of the standard techniques of reduction in classical genetics, can proceed. It appears, therefore, that what Wimsatt calls "identification," and what he correctly considers to be a critical part of scientific strategy, must be occurring either in the formulation of (i) or its refinement to (i′). Yet, both of these are conditionals, and the game can be played further to introduce a statement of the form

(iv) "if δ is the part of chromosome Γ, then δ contains a sequence of DNA Σ"

and attempt further reductions exactly as envisaged by Wimsatt.

The transitivity of the conditional suffices for this procedure. Moreover, the use of conditionals has yet another advantage. It is routinely acknowledged even among biologists unconcerned with potentially spurious problems raised by philosophical formalism that the molecular definition of a gene, which would have the form of a biconditional, is notoriously difficult (and difficult because of empirical complexities; see Chapter 6, § 6.5). However, the difficulty is not in finding acceptable sentences of the form (i), (i′), or (iv) – genetical investigations routinely provide such statements. Rather, the difficulty is in finding acceptable converses, sentences of the form, "if a is a DNA sequence with the set of properties Θ, then α is a gene," the difficulty being in the precise specification of Θ. This problem will be discussed in more detail in Chapter 6, § 6.5. Suffice it here to note that the pursuit of explanation in molecular genetics does not normally require these converses; forms such as (i′) or (iii) are good enough in scientific contexts.

If these problems, which have arisen from requiring biconditionals in the connections, are not sufficient to generate more tolerance, there are additional (logical) problems that have long been known. For instance, Kemeny and Oppenheim (1956), in their criticism of the Nagel-Woodger model, assume that the connections have the form of biconditionals and argue that once Nagel's condition of connectability is satisfied, there is no work left for the condition of derivability. Briefly, their argument is the following: the laws or theoretical statements of the reduced theory are presumably true. By postulation, the connecting statements are also true. Therefore, the replacement of terms in the former set of statements, using the latter, results in a set of true statements using only the terms of the reducing theory. These – assuming some unstated completeness (of deduction) criterion – must, therefore, be automatically derivable in the reducing theory. There are many flaws with this argument, including the assumption of deductive completeness, though Nagel accepts it and only reminds Kemeny and Oppenheim that the connections need not be biconditionals.[58] Almost by accident, Nagel captures the point about scientific practice that was made above.

Why, then, this apparent infatuation with the biconditional? Actually, it is an infatuation with identities and, more often than not, the ontological concerns that are connected with them.[59] Almost all those who present what amount as variants of the Nagel-Woodger or Schaffner model have such ontological concerns and proceed to impose even stronger restrictions on these connections.[60] In general the motivation seems to be what may be called "ontological reduction," that is, the potential elimination and replacement of the entities or predicates of the reduced theory by those of the reducing theory (see Chapter 3, § 3.8). The stronger restrictions (that is, those beyond requiring synthetic identities) are designed to connect entities and properties in the reduced theory (T_1) to those in the reducing theory (T_2) in some uniform or systematic way, for instance, by requiring that a single entity (or predicate) in T_1 cannot correspond to too many entities (or predicates) in T_2. Perhaps the strongest such restriction is the requirement that these synthetic identities connect "natural kinds" which, for both T_1 and T_2, are supposed to provide their basic ontologies (Fodor 1974). Unfortunately, even after several decades, there has emerged no uncontroversial explication of a "natural kind," what it is, or how, given a theory, one is even in principle, let alone in practice, supposed to separate natural kinds from other types of entities that occur in it.[61]

The usual conclusion in these analyses has been that reduction does not occur in whatever domain was being investigated. There are two reasons to be skeptical of these analyses: (i) if none of what are commonly held to be scientific reductions turn out to be such, there should be as much concern that the attempted analysis of reduction is faulty as any imperative to deny "reduction"; and (ii) it seems particularly unreasonable to expect, let alone to demand, that the description at the level of T_1, being a "coarse-grained" description (to borrow some terminology from statistical physics) should not be one in which a single term incorporates a variety of entities and processes at the "fine-grained" level of T_2. Moreover, if the point being made above, that the typical connection (and all that is required for explanation) is a conditional, then the question of identity become moot. Given the epistemological thrust of this book, issues connected with identity will not be pursued any further here.[62]

2.6. SUPERVENIENCE

Those whose concerns are primarily ontological often fail to find synthetic identities connecting terms and predicates that satisfy the other conditions that they may also impose. From this they sometimes proceed to deny that explanation is taking place and then, out of a customary respect for physical law, still maintain the primacy of physical law. Formally, such positions are negative, in the sense that they implicitly deny that the relation between the various phenomena can be captured by any formal model of explanation. As has been noted above, there is no inconsistency in adopting such a position: it involves giving a negative answer to the basic epistemological question of reduction, while either taking no position on or giving a (guarded) positive answer to some of the ontological questions.[63]

For instance, Fodor (1974) elaborates a possible position, "token physicalism," for the relation of mental to physical phenomena that makes no epistemological claim at all and a only very weak ontological claim. All that this position requires is that the mental phenomena do not violate physical law. In a more general context, it would require that the phenomena in the realm of the potential explanandum do not violate the laws from the realm of the (potential) explanans. Note how weak this requirement is: it only requires consistency, not even any semblance of determination from the latter realm to the former. The phenomena at the former realm could be entirely independent of the laws governing the latter.[64]

An ontologically stronger position that continues to have some adherents in the debates over reduction in genetics is based on the notion of "supervenience," as used by Davidson (1970) during a discussion of the relation of mental and physical events, and advocated in the biological context mainly by Rosenberg (1978, 1985, 1994). Davidson used "supervenience" to mean that "there cannot be two events alike in all physical respects but differing in some mental respect, or that an object cannot alter in some mental respect without altering in some physical respect. . . . Dependence or supervenience of this kind does not entail reducibility through law or definition" (1970, p. 88). Recalling the definition of "determinism" used in Chapter 1 (§ 1.4), supervenience of this kind entails the full determination of mental events by physical ones, but no explanation (if explanation is construed in any form involving laws) and not even the definition of mental events using physical laws (through identities). Moreover, the notion of supervenience need not be limited to the context of explanation through "law or definition" – one could argue for supervenience rather than explanation assuming any or even all models of explanation.[65]

Rosenberg (1978) has asserted this relation between Mendelian genetics and biochemistry: "the supervenience of Mendelian predicates on ... molecular ones enables us to identify states, conditions, events, and objects characterized by any of these terms with one another, without committing ourselves to the deducibility of Mendelian ... laws from the laws of bio-chemistry." Explanation is not possible in general – and Rosenberg (like Davidson) has deductive–nomological explanation in mind – because of complexity in the sense that apparently many lower-level (biochemical) states (or specifications) correspond to a single higher-level (Mendelian) state. In Chapter 6, § 6.6 it will be argued that, in molecular genetics, this suggestion of supervenience is a counsel of unnecessary despair and has its origin in an assumption of the necessity of having synthetic identities connecting terms and predicates. Wimsatt's model alone suffices to destroy the plausibility of supervenience as an adequate explication of what occurs in molecular genetics. If the demand for synthetic identities is dropped, the appeal of supervenience disappears completely in this, and probably many other, contexts.

3

Types of reduction: Substantive issues

This chapter will try to show that the formal issues that were discussed in the last chapter pale into insignificance – or, at least, into scientific and philosophical disinterest – in comparison with the substantive issues about reduction that arise once scientific explanations are considered in their full complexity. These issues are clustered around two questions:

(i) how is the system that is being studied [and the behavior of which is potentially being explained (or reduced)] *represented*?; and
(ii) what, exactly, has to be assumed about objects and their interactions for the *explanation* to work?

These questions have been posed generally enough to be applicable to all (natural) scientific contexts, including the physical sciences. Similarly, the analysis of reduction that is developed here is intended to be applicable to these other contexts. Nevertheless, given the limited scope of this book, the examples analyzed here will all be from genetics and molecular biology. Similarly, the general philosophical implications of this analysis that are drawn at the end of this chapter (§ 3.7–§ 3.12) will also be geared towards molecular biology and genetics though they are intended to be more generally applicable.

The basic strategy of this analysis will be to develop and use three substantive criteria to distinguish five different types of reduction. Three of these types of reduction are more important than others, and the rest of the book will proceed to use them to analyze the various types of explanation that are encountered in genetics. Two broad intuitions about reduction have guided the choice of the three criteria.

(i) Reduction involves the explanation of laws or phenomena in one realm by those in another. In this sense, reductions raise the potential for unification of knowledge, though this issue will turn out to be subtler –

and more controversial – than it may initially appear (see § 3.12). This intuition, that is, the requirement of two different realms, arises out of a desire to ensure that there remains some distinction between reductions and other kinds of explanation.

(ii) Many of the reductions will be attempts to explain properties of complex wholes in terms of their constituent parts. This was the kind of reduction that was expected of the mechanical philosophy of the seventeenth century, whether in the realm of physics or of the biological sciences. It was the motivation behind the formulation of the kinetic theory of matter, and the well-known attempts to reduce thermodynamics to the kinetic theory during the latter half of the last century. This is the type of reduction that is supposed to occur in molecular biology, when biological phenomena originally studied "classically" are explained on the basis of molecular mechanisms.

In the spirit of what was said in the last chapter, no assumption about the form of explanation will be made in the discussions of this chapter (or at any later point in this book). This is part of the general attempt to shift attention away from formal issues to substantive ones. However, a few general assumptions about what any explanation must presume will be necessary for the discussions and will be explicitly stated in § 3.1. These will be kept as minimal and uncontroversial as possible and will be self-consciously formulated as substantive assumptions. Attention will then be focused on the additional criteria that must be satisfied by an explanation for it to be a reduction and, if it is a reduction, to be the type of reduction that it is. These additional criteria are the substantive criteria mentioned above. These criteria will also be formulated in a fashion general enough to be consistent at least with any of the models of explanation that are currently in vogue. In particular, they will be consistent with both deductive–nomological explanation and with the types of explanation that are based on Salmon's (1971) "statistical relevance" model. Consistency with the former is important, because its miscellaneous variants, together, continue to provide the most popular candidates for deterministic explanations. The types of explanation that emerge from modifications and extensions of the statistical relevance model, meanwhile, have much in their favor purely as a models of statistical explanation, even though their scope might not be quite as general as Salmon (1971) had initially claimed.

Thus, this analysis divorces the criteria for reduction from those for explanation in contrast to what occurs in the older analyses of reduction that were discussed in the last chapter. This strategy has two benefits: (i) it makes this analysis of reduction immune to specific criticisms of various models of explanation; and (ii) the separation has the additional virtue of encouraging a more concentrated focus on the precise nature of reduction than has been customary. For instance, a clear separation of issues connected to explanation in general from those that arise specifically in the context of reduction helps avoid problems about some models of reduction that are, ultimately, problems about explanation (in general). For example, Sarkar (1989) criticized Wimsatt's (1976b) model of reduction (discussed in the previous chapter, § 2.4) on the ground that it is counterintuitive because laws are treated as groups of individual facts. However, this is actually a general problem with Salmon's (1971) account of explanation, which Wimsatt adopts (with some modifcations, as noted in Chapter 2, § 2.5), rather than a problem that is specific to the ideas about reduction that Wimsatt introduces. Separating the issues clarifies this point. Moreover, Wimsatt's cryptic claim that some reductions involve "compositional redescription," which is a specific claim about reduction, is exactly the sort of issue that receives the further attention that it deserves, once the issues connected only with reduction are separated from those connected with explanation in general. The danger of this strategy adopted here is that even the minimal assumptions about explanation that are made may prove to be incompatible with some model of explanation. This cannot be ruled out a priori but, in the absence of a potential candidate model of explanation that raises such a problem, this possibility will be ignored.

3.1. EXPLANATION

To talk of reduction at all, some basic assumptions about "explanation" will be necessary. For the reasons indicated above, these will be kept as general and as mild as possible. There are four such assumptions.

(i) It will be assumed that an explanation begins with a *representation* of the system. The distinction between a system and a representation is important. What, in everyday language, would be called the "same system" can have more than one representation, depending on the context of investigation. A chromosome can be represented as a group of loci (as, for instance, in linkage analysis) or as a physical object (as in cell biology), depending on what the context (of investigation and explanation) is. A cell may be represented as a chemical system of a particular sort, or as a cybernetic system.[1] Obviously, when the same system has different representations, interesting questions about their mutual consistency may arise. These questions can be nontrivial. For instance, it was often suggested that the linear order of loci on chromosomes, as used in classical genetics, may not be consistent with the order of the physical parts of the chromosome that correspond to these loci; this point will be discussed in detail later in this book (see Chapter 5, § 5.5). Similarly, the choice of a representation is nontrivial; an explanation can fail because of a poor choice of representation. Representations are often indicated diagramatically. These pictorial representations will routinely be used in this book. Perhaps the most sophisticated type of these pictorial representations are the three-dimensional models – such as the double helix model of DNA – that have played a major role in the development of molecular biology. Finally, to emphasize what may be an obvious point, a representation need not be a description of a system in physical space. If **A, a** and **B, b** are each a pair of alleles at two different loci of a diploid organism in some population, **AABb**, **Aabb**, and so on are all adequate representations of genotypes of different individuals for the type of explanation that is attempted in classical genetics (see Chapter 5, § 5.4).

(ii) It will be assumed that what is being explained is some feature of a system as represented – a law it (fully accurately, the representation) obeys, an event in which it participates, and so on. Thus, the account here will be neutral about the role of laws, theories, or individual events as the explananda of a reduction.

(iii) It will be assumed that, given a representation, an explanation involves a process of scientific reasoning or argumentation that will generally be called a *derivation*. "Derivation" as used here must not be confused with the logician's notion of derivation – as, for instance, used in the Nagel-Woodger and Schaffner models discussed in the previous chapter. In Chapter 2, § 2.3, those were called deductions.[2] Derivations will have varying degrees of precision and mathematical rigor. In general, the degree and type of rigor that is appropriate depends on the scientific context. Mathematical rigor, for instance, is no virtue if to achieve that rigor, questionable assumptions must be introduced into the representation of the system. This point will be taken up in § 3.4. Some of these derivations (for instance, those that typically occur in molecular biology) will be relatively trivial. This will generally be the case in contexts where mathematical explanations are customarily not used.[3] In many of these instances, most of the explanatory work goes into the construction of the representation (or model). In many of these cases, the ultimate derivation may be no more than a verbal argument.

(iv) It will be assumed that any explanation uses a set of explanatory "factors" that are presumed to be the relevant ones.[4] These factors bear the "weight" of an explanation in the sense that they provide it with whatever force or insight that it has. If the explanation is to be put into deductive–nomological form, these are the factors that will be referred to in the general law that forms the basis for the explanation, that is, they are what makes the law *nomological*. If the explanation has the form of Salmon's list, then these are the factors used to partition the reference class into homogeneous subclasses. Using "factors," therefore, provides a neutral and unified term to refer to the relevant entities in radically different models of reduction.[5] The list of factors, as the explanation itself, is context-dependent. The context will determine what factors are relevant, that is, when explanations may stop and when they are incomplete.

These assumptions can, no doubt, be formalized further, though it is open to question whether such formalization would add any insight or would rather bias the discussion toward some particular class of formal models of explanation; this point will not be pursued here. In passing, it should be noted that assumptions (i) and (iii) will generally play the same role here as Nagel's conditions of connectability and derivability, respectively. That assumption (iii) will play such a role should be obvious. However, assumption (i) appears rather different, at least in form, from the condition of connectability. The reason that it generally plays the same role as that condition is that the representation of a system indicates how the system fits into the two realms that would potentially be connected through a reduction.

3.2. SUBSTANTIVE CRITERIA AND TYPES OF REDUCTION

With these assumptions about explanation, three criteria will be used to analyze and characterize different types of reduction. These criteria are substantive: they are about what assumptions are made during a (putative) reductionist explanation, rather than about the form that such an explanation may take. Briefly, the criteria are as follows.

(i) *Fundamentalism*: the explanation of a feature of a system invokes factors from a different realm (from that of the system, as represented) and the feature to be explained is a result only of the rules operative in that realm.

(ii) *Abstract hierarchy*:[6] the representation of the system has an explicit hierarchical organization, with the hierarchy constructed according to some independent criterion (that is, independent of the particular putative explanation), and the explanatory factors refer only to properties of entities at lower levels of the hierarchy.

(iii) *Spatial hierachy:*[7] the hierarchical structure referred to in (ii) is a hierarchy in physical space; that is, entities at lower levels of the hierarchy are spatial parts of entities at higher levels of the hierarchy. The independent criterion invoked in (ii) now becomes spatial containment.

These criteria will be discussed in detail in § 3.3, and § 3.5–§ 3.6. On the basis of these criteria, five different types of reduction can be distinguished.

(a) *Weak reduction*: only substantive criterion (i) is satisfied. A genetical example, which will be discussed in detail in Chapter 4, is the attempt to explain phenotypic features of an organism from a genetic basis using the properties of heritability.

(b) *Approximate abstract hierarchical reduction*: only substantive criterion (ii) is (fully) satisfied, whereas (i) is approximately satisfied.[8] This type of reduction arises from type (c) below, when the assumptions or approximations used in the derivation (of what is being reduced) cannot be fully justified from the rules operative in the more fundamental realm. (As will be discussed in § 3.3, the satisfaction of criterion (i) should be seen as a matter of degree.) This type of reduction is perhaps best seen as an intermediate step in the path toward a reduction of type (c). A genetical example is an explanation on the basis of linkage analysis (see Chapter 5, § 5.4), when not all the properties that have been assumed for the various loci and alleles involved can be fully justified. However, reductions of this type are rare in genetics and will not be considered any further in this book.

(c) *Abstract hierarchical reduction*: only substantive criteria (i) and (ii) are satisfied. Reduction in classical genetics is of this type.[9] The set of alleles and loci form a hierarchically structured genotype. The rules of genetics are assumed to be more fundamental than those governing the phenotype. However, this hierarchy is not necessarily embedded in physical space. This will be discussed in detail in Chapter 5, § 5.5. In the genetic context, that is, in the context of this book, this type of reduction will be called "genetic reduction."

(d) *Approximate strong reduction*: only substantive criteria (ii) and (iii) are (fully) satisfied, whereas (i) is approximately satisfied. This type of reduction arises from type (e) when, as in the case of type (b), the explanations involve approximations that cannot be fully justified. A genetical example (see § 3.6) is the use of "information"-based explanation in molecular genetics, even after it becomes clear that there is no fundamental theory of information transfer that can provide a basis for such explanations (Sarkar 1996). An even more important example is the use of the lock-and-key model of macromolecular interaction in explanations, if the reduction is being assumed to be a reduction to physics (or chemistry); see § 3.6 and Chapter 6, § 6.4.

(e) *Strong reduction*: all substantive criteria (i), (ii), and (iii) are satisfied. This is the type of reduction where the properties of wholes are explained from those of the parts. Note that when a move is made from types (b) and (d) of reduction to types (c) and (e), respectively, what is assumed as the fundamental realm may change. There is no constraint on what the interactions between the lower elements of an (abstract) hierarchy may be, as long as they are assumed to be more fundamental than those at higher levels. However, once the entities in the hierarchy become spatial parts, their interactions are defined by the known interactions of these spatial parts. In natural scientific contexts, these interactions will be physical interactions, where "physical" is to be construed broadly to include all chemical, macromolecular, and other such interactions from any of the physical sciences. This is the type of reduction that was involved in the mechanical philosophy, the kinetic theory of gases, and, as will be shown in Chapter 6, in many explanations in molecular biology. "Strong reduction," in natural scientific contexts (including that of this book) will also be called "physical reduction."[10]

Types (a), (c), and (e) are the most interesting types of reduction in the context of genetics. In general, it is open to question whether types (b) and (d), neither of which requires the full satisfaction of criterion (i) (that is, fundamentalism), should be regarded as types of reduction at all. Moreover, when the fundamentalist assumption fails, one could wonder whether the explanation at hand is, in fact, an explanation at all. These points will be taken up in detail in § 3.4.

Nickles (1973) was the first to make a distinction between what he called "domain-preserving" and "domain-combining" reductions which, in the classification given here, is basically a distinction between weak reduction [type (a)] and all the others. For domain-combining reductions, Nickles assumed that some variant of Schaffner's (1967b) model is appropriate. Domain-preserving reductions, according to Nickles, occur between a theory and its successor. For him, the succeeding theories get reduced to the preceding ones: special relativity, for instance, gets reduced to Newtonian mechanics when an appropriate limit is taken, such as the speed of light, $c \to \infty$. In such cases there is clearly no explanation of the more general reduced (special relativity in the example) theory by the less general reducing theory (Newtonian mechanics).

Such "reductions" obviously cannot be explanations – a preceding theory cannot explain its successor in any reasonable sense of explanation.

Moreover, this use of "reduction," which seems to have been borrowed from mathematics, is unusual in a scientific context.[11] Wimsatt (1976b) accepted Nickles' distinction and clarified it as a distinction between "intra-level" or "successional" reduction (Nickles' "domain-preserving" or "weak reduction" here) and "inter-level" reduction, by which Wimsatt meant what is being called "strong reduction" here. For interlevel reduction, Wimsatt offered the model that was discussed in Chapter 2, § 2.4. Wimsatt (1995) has something at least akin to the spatial hierarchy criterion [criterion (iii)] in mind when he refers to "material compositional" levels of organization. Since he recognizes other kinds of hierarchies, at least implicitly, he assumes the distinction made here between abstract hierarchical reductions of type (b) or (d) and spatial hierarchical reductions of type (c) or (e).[12] The classification developed in the preceding paragraphs captures these distinctions and offers a finer resolution of the types of reduction that may be separated on the basis of substantive criteria.

3.3. FUNDAMENTALISM

Reduction is pursued because of a belief that some other realm is more fundamental – that is, it can provide deeper understanding, can correlate disparate insights, and so forth – than the one that has been studied. It is necessarily a fundamentalist enterprise at least in this mild sense. This, rather than any sort of more ideological or ontological fundamentalism, is what the substantive criterion (i) tries to capture. It incorporates three distinct requirements:

(a) that a potential reduction draws its explanatory factors from a different realm;
(b) that the rules from that realm are, for some reason or other, considered to be more fundamental than those of the original realm; and
(c) what is to be explained can be derived from these rules using only fully justifiable logical, mathematical, or computational procedures.

In this analysis, these three requirements will not all be accorded the same status. The first two have to be met in order for criterion (i) to be approximately satisfied. Satisfaction of the third requirement is a matter of degree. If the first two requirements are met, but the third is not or is only met to a very limited extent, it will be said that the substantive criterion (i) is only approximately satisfied.

The asymmetry of status of the three requirements listed here needs some justification. The first requirement is necessary to distinguish reductions from any explanation: unless what is explained and what does the explaining come from different realms, all explanations will turn out to be at least weak reductions. However, in this type of reduction, there is some ambiguity about what "realm" can be taken to mean. [Once criterion (ii) is introduced, "realm" can be unambiguously specified as referring to a level in the hierarchy; see § 3.5.] Roughly, realms will be considered to be different if they are the domains explored by different research traditions. (Even special relativity will have a different realm, in this sense, than classical mechanics. Thus, it will make sense to talk of the reduction of classical mechanics to special relativity.) The rules operative in the more fundamental realm play, here, the role that the reducing theory (or branch) played in accounts of reduction such as those of Nagel or Schaffner.[13] In the discussion that follows, this realm will be called the "**F**-realm"; its rules will be collectively referred to as "**F**-rules"; and "**F**-justified" will mean fully justified on the basis of the **F**-rules.[14]

The second requirement is almost trivial in the sense that unless some such fundamentalist assumption is made, it is hard to see why a putative explanation would be an explanation at all. However, such an assumption was not introduced generally in explanations in § 3.1 in order to assume as little there as possible. Thus, that discussion permits the sort of explanation that involves telling a plausible story – for instance, the kind of story that is rampant in descriptions of evolutionary history or in other "historical explanations." However, reduction, as construed here, at least requires explanation of a somewhat stronger nature. That is why the second requirement has been made explicit, and also why that requirement, along with the first, must be met by every reduction.

The situation with the third requirement is rather different. If scientific explanations were always logical deductions of the sort that were required by the logical empiricists, or if they were even fully rigorous mathematical arguments with no implicit problematic assumptions, then the third requirement would be gratuitous. All explanations would then be naturally **F**-justified. However, especially in contexts where explanations bridge two different realms, approximations are endemic. Moreover, as will be seen in the next section, some types of approximations raise serious epistemological and interpretive problems. This has often been recognized, even in the context of physics, where, since arguments are usually in mathematical form, it is easier to trace and analyze approximations.[15] In particular, approximations may introduce factual assumptions about the system and thereby become, as Leggett (1987, p. 116) has put it, "more or less intelligent guesses, guided perhaps by experience with related systems." In circumstances such as these, **F**-justification will clearly be lost. To get a better grasp on this problem, a more systematic treatment of approximations will be attempted in the next section.

3.4. APPROXIMATIONS

On those rare occasions when logical empiricists and their analytic descendants have addressed the fact that approximations are endemic to scientific explanations, they have generally attempted to incorporate them into the deductive–nomological model to have "approximative D-N [deductive-nomological] explanation."[16] Usually the strategy has been similar to that in Schaffner's (1967) model of reduction, where what is reduced is a theory that has a strong resemblance to the original target of reduction. Discussions of this sort only avoid the problematic substantive issues raised by approximations which, in turn, require a careful treatment of the different types of approximations that are routinely made.[17] The following six sets of distinctions are designed to help such a classification. They almost certainly do not exhaust all the interesting distinctions about approximations that may be made, as they are designed specifically to help address questions that are pertinent to reduction.

(i) Approximations may be *explicit* or *implicit*. There are at least two standard strategies of implicit approximation.
    (a) The invocation of a customary procedure that implicitly makes an approximation. In classical genetics (for instance, in linkage analysis) it is standard to assume implicitly that crossing-over (that would lead to recombination) will not occur within a gene-specifying segment of DNA (see Chapter 5, § 5.4). This is innocuous in almost all contexts. However, there often are more problematic implicit assumptions in linkage analysis, for instance, that the penetrance of an allele – that is, the probability that it will have a recognizable effect – is equal to 1.0.
    (b) The invocation of a model or formula that makes such an approximation. The use of atomic models with spherical atoms and definite surfaces is perhaps the most routine example of such an approximation in molecular biology (see Chapter 6, § 6.2).
    There is much to be said for keeping approximations explicit: it makes it easier to gauge the effects of the approximation. But a stricture that all approximations should always be explicit would probably prove cumbersome in many explanatory contexts: the socialization of scientists guards against errors from most common implicit assumptions. In general, implicit approximations – like other implicit assumptions – are more likely to be made explicit when an explanation involving them runs into trouble. Implicit approximations are not necessarily problematic

for reduction. However, when an approximation is not **F**-justified, and this is not recognized because the approximation is implicit, there is a potential for mistakenly judging a reduction to be successful as well as for incorrectly classifying it by type.

(ii) Approximations may be *corrigible, incorrigible in practice*, or *incorrigible in principle*. Corrigibility is not to be construed absolutely. Rather, all that is required is the knowledge of some procedure for decreasing the effects of an approximation. Usually, this involves a procedure for introducing corrections. In classical genetics, the assumption of an infinite population is corrigible in principle and also in practice in many circumstances. The crucial steps were initiated by Fisher (for instance, 1930) and Wright (for instance, 1931) who investigated stochastic models of gene transmission, that is, models with a finite number of individuals in a population. Since reduction in classical genetics (see Chapter 5) uses rules from population genetics among its **F**-rules, the diverse strategies for the incorporation of population size into population genetics provide methods for correcting the approximation that the population size is infinite. However, these methods are cumbersome in all except the simplest (for instance, one locus) models of genetic influence.

In molecular genetics, assumptions about the size and shape of atoms within a macromolecule are incorrigible in principle if the **F**-rules that are used are those from quantum mechanics, which are necessary for a general account of chemical bonding, but which do not allow such atoms to exist (see Chapter 6, § 6.3). There is no general procedure for determining when an incorrigibility in practice reflects an incorrigibility in principle. In a particular context, however, it is often possible to make this judgment.

(iii) The maximal effects of an approximation may be *estimable, not estimable in practice*, or *not estimable in principle*. The question of estimability is different from that of corrigibility because even if the effect of an approximation can be estimated, for instance, up to an upper bound, it need not be removable. Conversely, an approximation may be partly corrigible without its (full) effect being estimable. The inability to estimate the effect of an approximation may be due to theoretical or experimental reasons. For instance, as will be seen in Chapter 4 (§ 4.6), the effect of assuming that the variance of a phenotypic character can be approximately represented as the sum of a genotypic and an environmental variance cannot be estimated in many experimental situations. The problem is particularly severe in the case of human populations where, for obvious ethical reasons, systematic breeding experiments cannot be carried out. While the latter problem may be regarded only as one of estimation in practice, the fact that not all ranges of possible genotypes and environments can ever be explored is a problem of estimation in principle.

(iv) Approximations may involve: (a) only *mathematically justified* procedures (such as taking limits); (b) only **F**-justified procedures; (c) both of these; or (d) neither.[18] **F**-justification is as strong a condition as mathematical consistency. Justification must come from prior factual commitments made on the basis of the **F**-rules, that is, not through the implicit introduction of new assumptions into a derivation. In population genetics the limit of an infinite population is both **F**- and mathematically justifiable. In molecular biology, if the **F**-realm is taken to be either quantum mechanics or quantum chemistry, the shapes attributed to the atoms are not **F**-justifiable.

(v) Approximations may be *context-dependent* or *context-independent*. Once again, this is a question of degree. In classical genetics, including segregation and linkage analysis, the number of alleles at any locus that are assumed to be relevant to a particular problem is a context-dependent approximation (and one that, though corrigible in principle, is usually not corrigible in practice). The usual procedure is to consider all easily distinguishable alleles as distinct and to lump all the others together as a single allele (the "normal" one). In molecular biology the assumptions about the behavior of water molecules (especially how they tend to form ordered structures) that are invoked to explain the hydrophobicity [see Tanford (1980)] are context-dependent approximations in the sense that water is not assumed to have exactly the same properties in other situations, for instance, in the discussion of ionic reactions (in solution).[19] Context-dependence is a useful heuristic for suspecting the lack of **F**-justification of an approximation. However, caution should be exercised in the use of this heuristic; for instance, there is no such justificatory problem for the context-dependent approximation in classical genetics when the number of alleles at a locus is set to one more than those that are easily distinguishable.

(vi) Approximations may involve *counterfactual* assumptions, or not. Throughout this book, "counterfactual" is to be construed simply as referring to assumptions not permitted by the **F**-rules that are assumed.[20] Counterfactual assumptions are endemic, though the extent to which they involve serious violations of **F**-rules is often hard to gauge. The ubiquity of counterfactual assumptions raises the obvious ontological question of the status of results (such as the existence of certain processes) obtained using them. One response would be to assume that all theories are approximations, and that the "underlying world" poses no problem – in effect, use instrumentalism to rescue realism. Another is to admit that these counterfactual assumptions should be regarded as new factual hypotheses of the special sort indicated in the text. This would already raise problems about whether the fundamentalism criterion for reductions can even be approximately satisfied but, perhaps worse, this raises even more serious conceptual difficulties. What kind of factual hypothesis is it that allows the number of loci or alleles to be infinite, especially in a finite population? At the very least, further distinctions between types of counterfactual approximations will be necessary. This point will not be pursued here. Rather, the position that will be adopted is that counterfactual approximations, however they are interpreted, pose problems for judging the success and classificatory status of reductions.[21]

If an approximation, preferably explicit, is corrigible, its effects estimable, both F- and mathematically justified, context-independent, and involves no counterfactual assumption, it is presumably philosophically unproblematic, whether one is interested in only epistemological questions or also in ontological ones. As has previously been noted, too many philosophers who have analyzed reduction have assumed that derivations involve no approximation, or only approximations of this sort, in which case they can potentially be removed to recover, at least partly, the logical cleanliness that these philosophers value. Note that even in this situation, the use of approximations does not allow at least one conclusion that the traditional accounts of reduction assume to be true, namely, that reductions are transitive.[22] A sequence of approximations, however justifiable each may be by itself, need not be so justifiable.

Scientific developments rarely proceed according to the strictures on approximations imposed at the beginning of the last paragraph. Most approximations violate many of these strictures. As has been noted before, that an approximation is implicit is not necessarily problematic, and not much that is sensible can be said about the problem of counterfactual assumptions in general. There is also no general procedure for systematically judging the extent of the problems posed by the other violations of those strictures, that is, by violations of corrigibility, effect estimability, mathematical justification, and F-justification, either individually or jointly. Suffice it, here, to note only five general points about these approximations, the first four of which are generally relevant to explanations and emphasize the fact that approximations are not always epistemologically devastating. The last, which concerns what is perhaps the most problematic type of approximation, is particularly troublesome in the context of reduction.

(i) Even if an approximation is incorrigible, an explanation may have force, especially if the data to be explained have more uncertainty than that induced by the approximation. In the genetics of natural populations, where accurate measurements are often impossible in practice, this is often the case.

(ii) If the effect of an incorrigible approximation can be fully estimated, the point made in (i) can be made even stronger to test whether the approximation leaves an explanation within the smear of the data. In classical genetics one could, for instance, show how much difference full or no dominance would make for a model that assumes nothing in particular about dominance (for an example, see Chapter 4, § 4.5).[23]

(iii) Even if the effects of an approximation cannot be estimated, an approximation may not be particularly problematic. It might, for example, be corrigible to a significant extent. Or, it might be both **F**- and mathematically justified and involve no counterfactual assumption, which would be reason enough to tolerate it.

(iv) Moreover, experience has shown that mathematically suspect procedures do not doom a scientific research program. More often, as with Galton's regression procedure for the study of continuous traits in populations, new mathematics can be constructed to rationalize what should be regarded as earlier heuristic procedures.

(v) However, if an approximation is not **F**-justifiable, there is necessarily a problem with the satisfaction of the fundamentalism criterion for reductions. In the discussions of this book, the degree to which that criterion will be judged to have been satisfied will largely depend on the extent to which the approximations (if any) that are used during a derivation (in a reduction) are **F**-justifiable. (These arguments, in the present context, will have to be qualitative.) In passing, it should be noted that while the lack of **F**-justification is philosophically troublesome, it may also open the way for scientific developments, as the **F**-rules may get refined or changed or new realms that can serve as **F**-realms get invented.[24] In fact, it is at least arguable that the ability to choose which type of violation of existing **F**-rules is likely to be scientifically fruitful, and which would not, is an important measure of scientific insight.[25]

3.5. HIERARCHICAL ORGANIZATION

It was explicitly assumed in § 3.1 that an explanation begins with a representation of a system. This representation may or may not be hierarchical. The abstract hierarchy substantive criterion (§ 3.2) of reductions requires that this representation be hierarchical with the system at the top of the hierarchy and other entities at lower levels; that the hierarchy be constructed in accordance with some "independent" criterion; and that the explanation of some feature of the system refer only to factors at lower levels. Using the terminology that has been developed since, the **F**-realm can only include entities at these lower levels of the hierarchy, and the **F**-rules are those that they follow.[26] That the hierarchy be constructed according to an "independent" criterion means that it should not have been posited only for the sake of the explanation at hand, that is, there must be some independent reason for constructing it. There are at least three ways in which this independence condition could be satisfied (these are not independent): (i) there could have been other explanations (perhaps even of similar phenomena) that used such a construction, as was the situation with seeking Mendelian explanations around 1900, when it was not clear how the Mendelian factors would correspond to any physical entities of organisms; (ii) the hierarchy is constructed according to the dictates of a general research program such as the search for physical explanations in biology; (iii) the same hierarchy is used for some entirely different type of explanation. For instance, the genetical hierarchy can be used to explain the details of the origin of a complex trait (gene expression), while the independent reason for constructing it could be provided by the transmission of the trait. In the examples of genetic reduction discussed in Chapter 5, this is precisely the explanatory strategy that will be followed.

A standard way to represent such a hierarchical explanation is by a directed graph with at least one sink, no cycles, and with the edges being those admitted by the hierarchy, and their direction being determined by the direction of explanation.[27] Since not all entities of a hierarchically organized system will have to be invoked in every explanation, this graph will only select those entities whose interactions are relevant. (It can, therefore, be embedded in a larger graph representing the entire hierarchy.) One of the sinks in the graph represents the system whose features are being potentially explained. The "level" of an entity is the number of edges from that sink to the vertex representing the entity. The higher that this number is, the lower the level (see Figure 3.5.1).

Particularly simple hierarchies can be represented as trees, where there is only one sink, and every other node has a unique ancestor and any number of descendants (including 0). A typical hierarchical structure encountered

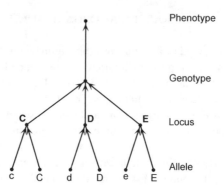

Figure 3.5.1. Graph Structure of RH Blood Group (in humans). Three (closely linked) loci, **C**, **D**, and **E**, determine the RH phenotype of an individual. It is usually assumed that there are two alleles at each locus: these are symbolized c or C, d or D, and e or E. The graph shown is for an individual with a *cde/cDe* genotype. The individual is only heterozygous for the **D**-locus. The arrows are in the direction of explanation.

in genetics is shown in Figure 3.5.1. A phenotypic feature (of humans) is the *Rhesus* (RH) blood group (Vogel and Motulsky 1986). The particular group that an individual has is explained on the basis of three loci, and the two (of many possible) alleles at each locus. This graph has a simple tree structure. However, should a locus be pleiotropic, and several features be studied simultaneously (say, as a complex trait), then the tree structure may be lost even in a genetic context. An example, represented in Figure 3.5.2, is a morphological abnormality (called the podoptera effect) in *Drosophila melanogaster* that is explained by abnormalities in both the wings and the legs (Goldschmidt 1955). These both seem to arise from a mutation at a single locus (with two alleles). The explanatory graph is no longer a tree.

Can the hierarchy criterion fail in biology? Sometimes, though not in the usual circumstances encountered in genetics. It fails in those evolutionary explanations that invoke fitness and, rather than postulating fitnesses as primitive properties of entities, attempts to explain fitnesses in terms of higher-level entities such as the (ecological) environment of the entities.

Note that directed graphs of this sort can represent any (abstract) hierarchy, and this representation makes no commitment to any entities or processes in physical space. Explanations represented in this way reflect intuitions about reduction because there is a definite hierarchy, allowing the assignment of "levels" to which different factors belong, and the requirement

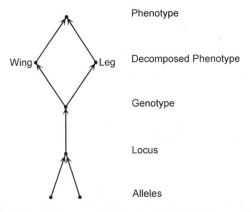

Figure 3.5.2. Graph Structure of Podoptera Effect (in *Drosophila melano-gaster*). The phenotype is subdivided into two separate phenotypic characters, a wing abnormality and a leg abnormality. They are both caused by the genotypic character of a single locus with two alleles. The arrows are in the direction of explanation.

that there be no cycles ensures that explanations proceed in a definite direction, that is, intuitively from lower through higher levels toward a sink. If an explanation only fully satisfies the hierarchy criterion – that is, it is representable in this way, while barely satisfying the fundamentalism criterion – then it is a reduction of type (b), which is an approximate abstract hierarchical reduction. If the fundamentalism criterion is also fully (or, at least, to a great extent) satisfied, then the explanation is a reduction of type (c), an abstract hierarchical reduction. As noted before, distinguishing between type (b) and type (c) reductions does not add much insight to discussions of explanations in genetics. This distinction will, therefore, be ignored here and attention will be restricted to type (c).

3.6. WHOLES AND PARTS

Now suppose that the directed edges along one of these graphs not only represent the direction in which explanation must proceed, but also the relation "is a spatial part of." The hierarchy will then be said to be a "spatial heirarchy." The levels of the hierarchy are then usually called "levels of organization."[28] What this means is that the independent criterion by which the hierarchy is constructed is that of spatial containment. As in the general case of using graphs to represent the parts of the hierarchy that are relevant to a given explanation, not all spatial parts of entities at each such level must occur in the directed graph representing the explanation. The only ones that will be vertices of the graph are the ones that are relevant to an explanation.

Figure 3.6.1(a → f) is an abstract representation of the *lac* operon in *Escherichia coli*. It is intended, at a gross topological level, to reflect the actual three-dimensional structure of the system. A regulator locus or site (i) is responsible for the synthesis of a repressor molecule that binds to an operator locus or site (o) in the absence of the inducer molecule (e.g., lactose). Presumably because of steric hindrance, when the repressor molecule is bound to the operator locus, synthesis of lactose does not take place. In the presence of lactose, because of some interaction between the lactose and the repressor molecule, the latter can no longer bind to the operator locus. In such a circumstance, β-galactosidase, which digests lactose, can be produced through the usual cellular transcription and translation processes. Another protein that is also produced is a permease molecule that aids the transport of lactose across the cellular membrane. The function in lactose digestion of a third protein, acetylase, if any such function is present, has not yet been deciphered. Figure 3.6.2(a,b) is the graph-theoretic representation of the explanation of gene regulation by the operon model.

This full significance of this example will become clearer in Chapter 6, § 6.2. Suffice it here to note six points that are particularly relevant to the present context.

(i) Once spatial representations such as Figure 3.6.1 are available, the graph-theoretic representations may not be particularly informative, at least in relatively simple cases. Graph-theoretic representations will be dropped in any further consideration of strong reductions in this book. Their utility is generally limited to situations when one needs a neutral way to represent what are abstract (nonspatial) hierarchies.

(ii) The abstract representation in Figure 3.6.1 is supposed to reflect actual steric properties. Most importantly, it reflects the belief that steric lock-and-key fits are the critical mechanism by which the various biological interactions are mediated. This point will be taken up in Chapter 6, § 6.2 and § 6.3. Ultimately, the abstract representation would be replaced by an actual model with (models of) the relevant atoms in place. This model is usually constructed by crystallizing the molecular complexes and solving the crystal structure, which is a laborious task.

(iii) The graph-theoretic hierarchical representation of the system, though obviously dependent on the spatial relations in the system, is not visually isomorphic to the three-dimensional structure of the system.

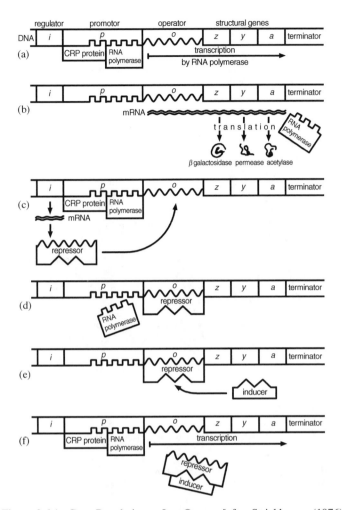

Figure 3.6.1. Gene Regulation at Lac Operon [after Strickberger (1976) p. 680; reprinted by permission of Prentice-Hall, Inc., Upper Saddle River, NJ]. (a) In the absence of a repressor protein [see (c)], and in the presence of CRP-protein attached to part of the promoter site (p), RNA polymerase also attaches to p, and transcription begins; (b) subsequently, translation takes place, and the proteins β-galactosidase, permease, and acetylase are produced; (c) transcription from the regulator site i and subsequent translation produces a repressor protein molecule that attaches to the operator site (o); (d) attachment of the repressor protein to o prevents transcription through (steric) hindrance. It is possible that it also prevents the attachment of RNA polymerase to p through the same mechanism. (e) An inducer attaches to the repressor and the complex detaches from o; see (f). This takes the system to the state described in (a).

(a)

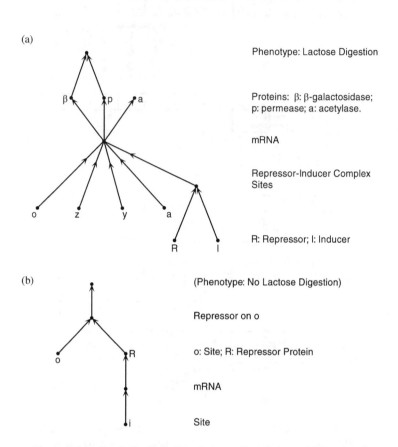

Phenotype: Lactose Digestion

Proteins: β: β-galactosidase;
p: permease; a: acetylase.

mRNA

Repressor-Inducer Complex
Sites

R: Repressor; I: Inducer

(b)

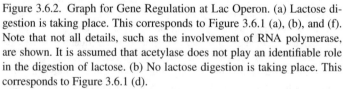

(Phenotype: No Lactose Digestion)

Repressor on o

o: Site; R: Repressor Protein

mRNA

Site

Figure 3.6.2. Graph for Gene Regulation at Lac Operon. (a) Lactose digestion is taking place. This corresponds to Figure 3.6.1 (a), (b), and (f). Note that not all details, such as the involvement of RNA polymerase, are shown. It is assumed that acetylase does not play an identifiable role in the digestion of lactose. (b) No lactose digestion is taking place. This corresponds to Figure 3.6.1 (d).

(iv) There is another kind of failure of isomorphism that is perhaps even more interesting. The "natural levels" determined by the spatial hierarchy of organisms (organism → organ → tissue → cell → organelle → macromolecule → molecular moiety, and so forth) are not usually the same as the levels in the graph representing the reduction (which, by definition, is what is being called "level" here). The explanation of the behavior of the cell may involve organelles, macromolecules, and solvent particles. This point, which emerges naturally from the discussion of reduction given here, has sometimes been systematically elaborated as the claim that reductions can require interfield theories.[29] However, this point is nontrivial if (and probably only if) "reduction" is construed according to the Nagel- or Schaffner-type models.

(v) Note that the type of strong reduction this explanation provides is incomplete in the sense that some parts of the model, such as the appeal to steric hindrance, are no more than conjectures at this stage. However, they are valuable as heuristics in the sense that they provide specific targets for future inquiry.

(vi) The explanation is incomplete in yet another way insofar as one can ask why steric hindrance occurs. After all, this seems to have been a principle gathered from mechanical models built with everyday objects rather than from fundamental physical principles. It is also possible that this incompleteness will generate further inquiry. However, historically, this has not been the case in molecular biology. This point will be briefly addressed below and in more detail in Chapter 6.

Note that the satisfaction of the physical hierarchy regulatory criterion requires the satisfaction of the abstract hierarchy criterion. If an explanation fully satisfies only these two criteria and approximately satisfies the fundamentalism criterion, it is a reduction of type (d), that is, an approximate strong reduction. If it fully (or to a very large extent) satisfies the fundamentalism criterion, it is a reduction of type (e), that is, a strong reduction. Strong reductions have a long and illustrious history, from the mechanical philosophy through the kinetic theory of gases at the end of the nineteenth century to contemporary molecular biology.

However, whether reductionist explanations in molecular biology are actually strong reductions or only approximate strong reductions [of type (d)] is a rather subtle question.[30] The answer depends critically on what is taken to be the F-realm. If the F-realm is taken to be the physics and chemistry of macromolecules, where the F-rules have been determined empirically, then it is plausible to argue that strong reductions are taking place. This is what was implicitly assumed in the discussion of the operon model that was given above. However, there is something epistemologically unsatisfactory about such a move. The physics and chemistry of macromolecules, at least at this point in the development of science, is not a particularly well-developed or organized domain of inquiry. At best, one can say that there are some experimentally gleaned rules (such as the importance of steric hindrance) that define it. Moreover, at present these rules cannot be reduced to any better-defined (and more fundamental, in this context) realm such as physics and chemistry. Not only have systematic attempts in this direction been very rare, but also the types of approximation involved suffer from all the problems noted in § 3.4; this point will be developed in Chapter 6, § 6.3. Moreover, it is far from clear that the approximations that seem to be necessary in one context would be consistent with those needed in another. In most situations the posited macromolecular interactions assume a spherical water molecule. But for water to have the internal structure to account for the hydrophobic interaction, it cannot be spherical. There is no glaring inconsistency here, but sufficient inconsistency at the level of fine detail to have some reason for worry, if not for serious skepticism.

There is a temptation, therefore, to ask for a reduction where the F-realm is fundamental physics or chemistry. The motivation is that most of the interactions in the molecular realm are known to be mediated by quantum mechanics, as in the case of usual (covalent) chemical bonding. The trouble is that the fact noted at the end of the previous paragraph, that there is no epistemologically respectable way to carry out derivations from this realm to the experimental rules discovered for macromolecules, precludes any easy direct reduction of biological phenomena to such a quantum F-realm. (Of course, even if the former reduction was possible, it would not guarantee that the total reduction could be accomplished. The "goodness" of approximation is not always a transitive relation, as was noted in § 3.4.) Why has this situation not generated further work? The answer simply seems to be that the usual models of molecular biology, with all the approximations and possible inconsistencies built into them, work very well at the level of experimental detail that is currently available.[31] This is the beauty – and puzzle – of molecular biology from a philosophical point of view. Further discussion of these points will be found in Chapter 6.

3.7. EPISTEMOLOGICAL ELIMINATIVISM

The last six sections of this chapter will very briefly discuss some general philosophical issues that are affected by what is decided about reduction. These discussions are less than comprehensive in two ways: (i) given the restricted scope of this book, only those issues that are interesting in the context of genetics will be treated; (ii) even when an issue is discussed, those aspects of it that are not particularly interesting in a genetic context will generally be ignored. This is particularly true of ontological questions that are far less pressing in the case of genetics than, for instance, in the case of neurobiology.

This section will reiterate a point about epistemological eliminativism that was already made in the previous chapter (§ 2.4). It has sometimes been thought (for examples, see § 3.12) that reductions or, at least, reductions of type (a), (c), or (e) would lead to the disposal of the reduced rules, theories, and so forth, in favor of those that provide the reductions. These would be appropriately called "reductions with replacements." This is a strong thesis of epistemological eliminativism: the reduced entities would be entirely dispensed with because of reduction. Thus, special relativity would replace Newtonian mechanics, the kinetic theory of gases would replace the thermodynamics (of gases) and, rules about genes would replace those about phenotypic features. There is little to be said in defense of this strong thesis. In fact, from Hull (1972, 1974) on genetics, to P. Churchland (1979, 1981, 1984) and P. S. Churchland (1986) on folk psychology, replacement is usually presented as the type of scientific change that occurs when reduction does not take place. In fact, as modified by Sarkar (1989), Wimsatt's (1976b) model of reduction provides an explicit rationale for not abandoning the laws or theories of the realm being reduced (see Chapter 2, § 2.4).

There are at least two reasons why reduction with replacement is an unlikely possibility.

(i) All known cases of reduction show that using only the rules from the F-realm would lead to much more complex explanations than tolerating the rules from the reduced realm.[32] This was the important point that Wimsatt (1976b) made. It is because of this fact that in biology, cell biology and organismic biology continue to be used with much of their traditional conceptual apparatus, in spite of many partial reductions to molecular biology (see Chapter 6).

(ii) In fact, in many cases reductions generate further confidence in the use of rules from the reduced realm because they explicate the exact range of applicability of those rules.[33] The success of special relativity, and the reduction of Newtonian mechanics to it, shows that Newtonian mechanics can be used without worry at low velocities. Similarly, Newton's law of gravitation remains adequate for the strength of gravitational interactions experienced on earth even though it can be reduced to general relativity. Even dubious reductions involving many problematic approximations can serve this role. A genetic example, Fisher's reduction of biometry to Mendelian genetics, will be discussed in detail in Chapter 5, § 5.2.

There is a weaker thesis of epistemological eliminativism that is more interesting. This would say that, although rules from the reduced realm continue to be used for pragmatic reasons, they receive their epistemological force from the **F**-rules or, in other words, the weight of the explanation is still borne by the **F**-realm. Many different versions of this thesis, flushing out different senses of explanatory weight, can easily be generated. However, there is a compelling general argument against them. If reductions never involved problematic approximations, this thesis could probably be maintained. The approximations that are obviously most problematic are those that are context-dependent approximations and those that cannot be either **F**- or mathematically justified. But even other ones, including those that involve counterfactual assumptions, are reason for worry. These worries are particularly troublesome if the reduced rules have strong empirical support. If, indeed, their support is significantly stronger than that of the **F**-rules, then, along with Cartwright (1983), one could even worry about the correctness of the **F**-rules in the presence of such approximations. Moreover, once all these approximations are banished, very few reductions remain.[34] Therefore, at best, even this weak thesis can only be very occasionally maintained.

3.8. ONTOLOGICAL ELIMINATIVISM

Like epistemological eliminativism, ontological eliminativism comes in two versions, one strong and rather absurd, and the other weak and occasionally defensible. The strong version would replace all entities and processes from the reduced realm with those from the **F**-realm following a reduction. It is hard to find anyone who explicitly advocates such a position.[35] Nevertheless, unless this result is the goal, there is no rationale for the obsession with synthetic identities that was noted in the previous chapter (§ 2.5). All other possible ends of reduction, including the two types of epistemological eliminativism and the weaker form of ontological eliminativism, do not even require biconditionals, let alone identities, as the relation between the entities and processes of the **F**- and other realms. Conditional statements would suffice for these purposes.

The strong form of ontological eliminativism is impossible to sustain. The problem, once again, is a consequence of the ubiquitous approximations that lead to incompatible properties routinely being ascribed in different realms to what is supposed to be the same entity (or process). It is hard to imagine, unless one suspends all concern for conceptual rigor, how the atom admitted by quantum mechanics can replace all the atoms of chemistry, let alone molecular biology.[36] The attempt to do so, like many other forms of ontological fundamentalism, could generate interesting scientific projects: a devoted quantum mechanical atom-fundamentalist could do good science trying to carry out such an elimination. Strong ontological convictions, however unjustifiable they always seem in retrospect, have the obvious virtue of quite often being able to generate worthwhile research programs. But that is a different question from that of their actual viability in practice.

A weaker form of ontological eliminativism accepts all the problems raised by approximations, and so on, but nevertheless asserts that all entities and processes occurring in some other realm are nothing but those occurring in the F-realm. In biology, with physics and chemistry providing the F-realm, this sort of eliminativism is perhaps most famously associated with Loeb's (1912) "mechanistic conception" of life. Positions that invoke this weak form of ontological eliminativism usually come with an associated epistemological program of attempting to carry out reductions with as few dubious approximations as possible. Unless such a program is associated with it, weak ontological eliminativism is a rather innocuous and vacuous thesis since there is no other way to spell out the meaning of the "nothing but" in its conception.

Nevertheless, even when there is an associated epistemological program, there are (at least) two ways to deny even the weak version of ontological eliminativism.

(i) One could deny it on ontological grounds and argue that new – or, at least, slightly different – entities and processes exist in different realms. (This would, among other things, explain why epistemologically questionable assumptions are necessary when traversing realms.) Traditional Marxism – or, at least, its dubious official metaphysics, dialectical materialism – required a move of this sort: each level or organization of matter had its own (obviously "emergent") laws and so on.[37] So did vitalism and other such moribund doctrines that probably reflect little more than emotional discomfort with the anemic ontology of the eliminativists.[38]

(ii) Or, one could deny it on conceptual and methodological grounds. The point here is that in the absence of any reasonably consistent scheme that encompasses the different realms of inquiry, it is hard to see how one can make firm (let alone final) ontological commitments about what an electron, atom, or fitness is. This is the conceptual ground for denying even weak ontological eliminativism. The methodological ground is that it is equally hard to see – unless one is obviously and deeply influenced by some extreme (e.g., Quinean) form of philosophical fundamentalism – why one would want to suggest that entities and processes in the reduced realm are "nothing but" those in the **F**-realm. Commitments of this sort are not necessary to pursue science – to construct useful or intellectually illuminating models of the world. By this point, no reader should be surprised that this is the option that will be endorsed in this book. (There is, no doubt, something reminiscent of logical empiricism in this move; to some extent it captures part of what was valuable in that school's resolute refusal to concern itself with ontology.)

To reiterate the position being advocated here (in this section and the previous), to say, for instance, that a biological system is "nothing but" a physical system has only one useful consequence: It might lead to an epistemological program of the physical investigation of that system. But one can proceed with such an investigation without any such eliminativist move. In fact, as noted in Chapter 2, § 2.2, and as discussed again in Chapter 6, § 6.1, Delbrück, following Bohr, helped initiate molecular biology by pursuing physical explanations hoping that they would ultimately fail.[39] Thus, not only was ontological eliminativism not endorsed, but even epistemological eliminativism was rejected.

3.9. REDUCTION VERSUS CONSTRUCTION

Those who doubt the possibility of reduction sometimes raise the point that, whereas one might find a reductionist explanation of some phenomenon after it has been described in its own realm, one would not have been able to suggest (or in a weaker version would not, in practice, have suggested) such a phenomenon if one only had access to what is known about the **F**-realm. This observation is clearly correct for many cases but it should not be interpreted to be more important than it is. Rather, three points should be noted about it.

(i) The issue here is partly one about the difference between prediction and explanation. At best what this criticism points out is that one would not have predicted all the consequences of some set of assumptions (even if all the rules were fully deterministic). Explanation is usually a weaker category than prediction.[40] Reduction, being a type of explanation, may not always permit prediction. However, this is a general problem about explanation; it does not provide any compelling reason to doubt the value of reductions.

(ii) If either physics or biology is a guide, what happens in any realm depends on both the entities and processes and the (boundary or initial) conditions of the entities on which those processes act. In biology, those conditions are often critical in the explanation of phenomena – this is one version of what is sometimes called the principle of historicity in biology. The entities and processes are clearly not sufficient to construct the outcomes. Moreover, since the set of possible conditions in which the entities may find themselves is large, it should come as no surprise that biological outcomes – the result of a particular evolutionary and a particular developmental history – would not be predicted or constructed (from some other **F**-realm) in practice. But this, once again, is no argument against the value of reductions.

(iii) Finally, and this is probably the most important point to be made in this context, approximations are not only generally intransitive (as has already been noted in § 3.4), but they are also often degenerate in the sense that there is no unique approximation that is the only correct one in a particular context. When approximations are even to a slight degree context-dependent, there is little chance that the correct approximation can be chosen without reference to what is being explained.

3.10.   REDUCTION AND SCIENTIFIC METHOD

The discussion in this book has so far been focused on the questions of whether certain explanations are reductions and, if they are, what type of reduction subsumes them. Yet, there is an entirely different question about the role of reduction in science, namely, the part that it plays in methodology. One aspect of this, reduction as a research strategy, has often been studied.[41] The basic idea here is that research strategies could be designed to search for reductions. How reduction is construed is critically important in this context. If it is construed according to Schaffner's (1967b, 1993a) model, then as Schaffner (1974, 1993a) has noted, reduction is "peripheral" to the practice of science (at least in the case of molecular biology). If, however, reduction is construed according to Wimsatt's (1976b) model, a better case may be made for the claim that the pursuit of reductions is central to many scientific research programs. Though this point will not be pursued at length here – there will be many examples in later chapters – each of the three important types of reduction [types (a), (c), and (e)] that were distinguished above has been actively pursued in genetics.[42] Liberated from the formal models, the value of reduction as a research strategy in genetics cannot be seriously denied.

Suffice it here only to note that in physics the search for strong reductions was the critical motivation for Maxwell, Boltzmann, and other physicists who pursued mechanistic explanations for the laws of thermodynamics in the latter half of the nineteenth century. In contemporary behavioral and psychiatric genetics the search for reductions, especially through linkage analysis, is the dominant research strategy; this point will be elaborated in Chapter 5, § 5.4. Perhaps the most interesting – and, in many ways, odd – use of reductionist research strategies was in some of the work of the Phage Group in molecular biology in the 1940s, when the reduction of biology to physics was pursued even when some members, including Delbrück and Stent, were expecting and hoping that reduction would ultimately fail.[43] Such a possibility, of course, illustrates how the pursuit of reduction as a research strategy is a different issue from whether a given explanation is reductionist. Reductionist research programs may fail to provide explanations at all (as will be argued for one kind of attempt to reduce phenotypic traits to genotypic ones in Chapter 4), or they may come up with perfectly respectable explanations that are not reductions (of whatever type is desired).

A second and more important role for reduction in methodology arises from the fact that successful abstract hierarchical and strong reductions [of types (c) and (e)] are routinely used to generate investigative tools where the continued success of new reductions is assumed and used to chart out a domain. In the abstract, the point is this: reductions of these types assume a particular hierarchical structure in the representation of a system. If it is assumed that a particular feature can be reduced according to such a pattern, then, from the existence of the feature, the existence of these internal structural features can be inferred. For instance, the reduction of some phenotypic traits to genetics between 1900 and 1912 generated the programs of segregation and, especially, linkage analysis, which assumed that such reductions would occur for other traits, and then used this assumption to map traits to factors and, in the case of linkage analysis, to specific linkage sets. (These processes will be analyzed in more detail in Chapter 5, § 5.3 and § 5.4.) Further, when these linkage sets were interpreted as corresponding to chromosomes, this procedure led to the systematic mapping of genes (loci) onto specific regions of chromosomes, starting with the work of the Morgan group at Columbia University in the 1910s and 1920s and continuing to this day.[44]

3.11.   THE VALUE OF REDUCTIONS

Chapter 2 was quite critical of Nagel's and other formal models of reduction that regarded it necessarily as a relation between theories. However, unlike Schaffner (1967b, 1993a) and most other proponents of formal models of reduction, Nagel (1961) did not limit his attention to formal issues but coupled his formal model to a discussion of nonformal issues about the value of reductions in scientific practice. That discussion is characteristically illuminating. The formal model, Nagel correctly noted, "[did] not suffice to distinguish trivial from noteworthy scientific achievements."[45] Reductions had to satisfy at least three conditions to be of significance.

(i)   The assumptions of the reducing theory (its laws and theoretical postulates) should not be ad hoc in the sense of having been introduced only to carry out a particular reduction. Rather, they must be "supported by empirical evidence possessing some degree of probative force" (p. 358).

(ii) The reducing theory "must also be fertile in usable suggestions for developing the [reduced theory], and must yield theorems referring to the latter's subject matter which augment or correct its currently accepted body of laws" (p. 360).

(iii) The reduced and reducing theories must be at a stage of their development where such a reduction aids the development of the reduced theory and, at the very least, does not frustrate its development by shifting interest away from it to the reduction or the reducing theory.

The continual references to theories (and to theorems) is no more than an expected artifact of Nagel's formal model. However, once those references are removed and replaced by the references to the different realms under consideration, little needs to be added to Nagel's account. Nagel shows no tendency to eliminativism of either kind, has a good appreciation of the scientific process as being dynamic, and has a healthy respect for the pragmatic component in the evaluation of scientific developments. Perhaps most importantly, he does not offer a blanket endorsement of reductionism, that is, of the thesis that (ultimately) reductions are bound to be successful in a given field. As the discussion in Chapter 6, especially § 6.6, will show, part of the defense of (strong) reductionism in molecular genetics will be to point out that attempts at reduction have been remarkably fertile in generating fruitful fields of inquiry.[46]

The only one of Nagel's conditions that requires some systematic elaboration is the first. Nagel basically restricts what he calls an ad hoc assumption to what would, in the framework developed in this chapter, be called context-dependent assumptions (though not all context-dependent assumptions should be regarded as ad hoc) generally involved in approximations. However, these are not the only assumptions that would decrease the value of a reduction. As the discussion of § 3.4 showed, significant incorrigibility and an inability to estimate the effects of an approximation would both (independently) decrease the value of a reduction. Lack of mathematical justification could also be problematic and, of course, lack of **F**-justification may make a putative reduction fail to satisfy the fundamentalist criterion.

3.12.   THE UNITY OF SCIENCE

The demand that all science, if not all knowledge, be unified into a single structure has been a popular and recurrent feature of the Western philosophical tradition since at least the seventeenth century. When the logical empiricists attempted their reformulation of philosophy, which was heavily dependent on their interpretation of science, the unity of science served as one of their most important regulative principles. It is easy enough to see how systematic reduction could lead to the unity of science. As Quine ([1977] 1979, p. 169) puts it: "Causal explanations of psychology are to be sought in physiology, of physiology in biology, of biology in chemistry, and of chemistry in physics – in the elementary physical states."[47] If (and, presumably, only if) all these explanations {which would at least be weak reductions [type (a)]} are forthcoming, then science would be unified, at least with respect to epistemological concerns, with the elementary physical states and the rules governing them providing the unifying framework. This is Quine's version, or at least one of his versions, of physicalism.

In general, however, with Feigl (e.g., 1963) being perhaps the most notable exception, the logical empiricists did not endorse reduction as the route to the unity of science. Rather, the unity of science was to be achieved through what they called "physicalism," which was a rather different position than Quine's version (as described above). Physicalism went through several changes as the logical empiricist program unfolded – and, as some would say, disintegrated – and it was interpreted rather differently by the different figures associated with it, including Neurath and Carnap.[48] However, it was almost always the thesis that the same language be chosen to describe the experimental domains of all the sciences. Originally, this was the language of (presumably theoretical) physics, later it became the language describing everyday physical objects such as chairs and tables, and finally simply any nonsolipsist language (in the sense that it is a language in which all statements are intersubjectively confirmable (Carnap 1963).[49] Logical empiricism's demand for the unity of science is a rather innocuous doctrine; certainly, it makes no demand for any of the types of reduction being considered here.

Indeed, the thesis that the unity of science can be achieved through reduction seems only to have been clearly formulated by Oppenheim and Putnam (1958) in a manifesto written when logical empiricism was already on the wane. Oppenheim and Putnam have strong reductions in mind, and they assume that reductions involve the derivation of laws. They distinguish six levels of organization: those of elementary particles, atoms, molecules, cells, multicellular living organisms, and social groups. They are, of course, fully aware that strong reductions between all these levels were far from forthcoming; their thesis was intended as no more than a "working hypothesis." But, if that working hypothesis is supposed to describe all of scientific practice then, not only is it descriptively false but – as Fodor (1974) has argued – there are sound methodological reasons to doubt its utility.

Reductions can be unilluminating, as Nagel (1961) realized. Counterfactual, context-dependent, and other problematic assumptions made during reductions may not even allow weak epistemological eliminativism in many situations, let alone strong epistemological eliminativism (or, for that matter, any form of ontological eliminativism). At best, most reductions establish some (weak and not very precise) form of consistency between various realms; an explicit example of this kind will be treated in detail in Chapter 5, § 5.2. It will not be assumed in this book that all types of reduction necessarily contribute to unification. Approximate hierarchical or strong reductions [of types (b) and (d)] clearly do not.[50] However, those reductions that permit epistemological eliminativism do make such a contribution. But these may well be rare. Moreover, to the extent that reductions contribute to added confidence in the reduced theories, laws, etc., but involve problematic assumptions, they may well contribute to the disunity of science in practice.

Note, moreover, that one of the most successful unificatory theories in science, evolutionary theory, is manifestly nonreductionist [in any sense except perhaps (i)]. Even molecular biology sometimes makes use of nonreductionist modes of explanation, including functional explanation (which relies on evolutionary theory for its warrant),[51] and, as will be discussed in detail in Chapter 6, § 6.7, may even need what will be called "topological explanation." These examples point out that what may, at least intuitively, be called "deeper understanding" often requires nonreductive modes of explanation. Finally, no position will be taken here on the question of whether the unification of science is achievable or even desirable as a goal. Suffice it merely to note that while some generality seems to be required to distinguish an explanation from description in most circumstances, it does not follow that more generality alone guarantees better understanding in all circumstances. In general those who suggest the disunity of science and the need for special sciences will probably find many of the analyses of this book more congenial than their opponents.

4

The obsession with heritability

There is a form of reduction in genetics that attempts to explain phenotypic properties from a genotypic basis without attributing any particular structure to the genotype. Therefore, if successful such a reduction would be weak [that is, of type (a) from Chapter 3, § 3.2]. It consists of the use of concepts of "heritability" as measures of genetic influence.[1] This technique has been routinely invoked for many putative human traits including IQ (which has had a long and occasionally sordid history), vocational interests, "religiosity," "openness," "agreeableness," "conscientiousness," neuroticism, and extroversion.[2] The basic idea is that if some trait has a high heritability, its origin or, at the very least, why it varies from individual to individual can be explained from a genetic basis. The stronger claim is the more interesting one. Unfortunately, it is not even a distant approximation to the truth. However, the weaker claim is occasionally plausible. These are the two basic points that this chapter will make. These points are nothing new; however, this chapter will attempt to synthesize the past conceptual analyses of heritability.

The roots of heritability analysis go back to what was called "biometry," largely the work of Galton and, especially, Weldon and Pearson in the 1890s, which partly explains why it is relatively neutral with respect to the details of the structure of the genome. After Mendelism was recovered around 1900, starting with the work of Yule (1902) and ending with a classic paper by Fisher (1918), biometry was reduced to Mendelism. This reduction will be analyzed in detail in the next chapter (§ 5.2), where it will be shown that it involved a set of epistemically problematic approximations. At the very most, Fisher's "reduction" showed that biometry was consistent with Mendelism; it certainly did not provide new, specifically Mendelian, methods for the study of continuously varying traits. What Fisher did provide were even more powerful statistical methods than those that Pearson had devised. As a consequence, a statistical theory of "biometrical genetics" was developed that paid little more than lip service to Mendelism and eventually came to be called "quantitative genetics." Its main distinguishing feature – that is, what distinguished it from Mendelian (or classical) genetics – was its direct concern with continuously varying traits, such as the size or weight of organisms, rather than the discrete traits to which Mendel and his twentieth century followers have limited most of their attention.

In this context, in the early 1920s Wright (1920) and others began the implicit use of a concept of heritability. However, the two generally used concepts of heritability, "broad heritability" and "narrow heritability," were only explicitly distinguished and codified by Lush (1943) much later.[3] (A systematic account of these two concepts will be given in § 4.1.) "Heritability" thus crept into genetics largely unheralded as a theoretical concept; a detailed history of these developments is yet to be written.[4] For conceptual, methodological, as well as political reasons, the use of either of these concepts has remained controversial. As early as 1951, Fisher (1951) rejected the use of heritability on conceptual and methodological grounds. Nevertheless, heritability analysis became standard fare in quantitative genetics.

Broad heritability is supposed to differentiate between the relative contributions of genotypic (or genetic) and environmental influences in the genesis of differences in a trait, and to provide a measure of the former. Narrow heritability is less natural. In the phenogenesis of a trait, it is supposed to differentiate between everything else and the collective contribution of all alleles at all loci, when each allele acts additively (or completely independently) at its own locus (that is, there is no dominance) and with respect to all alleles at all other loci (that is, there is no epistasis). The quantitative value of narrow heritability defies direct experimental assessment. In practice, as the analysis of models in § 4.4 will show, narrow heritability is estimated by attributing as much influence as possible to additively acting alleles by curve-fitting.

This chapter will show that the aim of using heritability analysis to assess how much of the phenogenesis of a trait or differences in it can be genetically accounted for is not quite as easily realized as it is supposed to be. This point is well-known. Nevertheless, there is a long-standing tradition that not only interprets the concepts of heritability in this way in all contexts, but sometimes goes even further and interprets high values for heritability to be an indication of genetic determination. This tradition is not confined to psychologists, who have most often made such claims, or even to them and behavioral scientists. Sometimes, even biologists and philosophers, who are normally quite sensitive to the nuances of heritability, also make misleading statements. For instance, Sober and Wilson (1994, p. 538) claim:

> In biology, heredity is measured by the concept of heritability. When offspring phenotypically resemble their parents, this can be due to their sharing genes

or living in similar environments (or both). A phenotypic trait has nonzero (narrow) heritability when parent/offspring resemblance is attributable, at least in part, to shared genes. . . . As such, heritability is a property of the phenotypic traits of organisms.

The purpose of this chapter is to examine such uses of heritability (and also to analyze how it fits into the general context of genetic reductionism).

For such ambitious interpretations of heritability to be valid, two criteria of the following sort must be satisfied. The first of these has a strong and a weak version:

(i) the heritability of a trait is a monotonic function of the degree to which that trait can be accounted for from a genetic basis;

or, at the very least,

(i′) the heritability of a trait has a high value (on an appropriate scale) if and only if that trait can be accounted for from a genetic basis.[5]

If even (i′) is not satisfied, there is obviously no justification for regarding a high value of heritability to be indicative of anything at the genetic level. However, there is good reason to want (i), which is stronger. Without (i) a low heritability need not be interpreted as being indicative of influences from a nongenetic (environmental) level. In such a situation claims such as those of Sober and Wilson (1994), who apparently want hereditary and, therefore, genetic influence to be "measured" by heritability could not be sustained. Therefore, in some sense, a genetic explanation of the origin of the trait would still remain somewhat wanting.

The second criterion is even simpler:

(ii) the heritability of a trait of an individual should only be a function of its genotype and its developmental (that is, nongenetic) environment.

The main purpose of this criterion is to ensure that the heritability of a trait remains a property of an individual and is not *irreducibly* that of a population.[6] "Irreducible," here, is being used to distinguish between two possible types of property: (i) a property that is initially or naturally defined for a population but can be defined for individuals without reference to other members of the population; and (ii) a property that is not only initially or naturally defined for a population but cannot in principle be defined for an individual in such a way. Only the latter type of property is "irreducibly" that of a population. This is a strong notion of irreducibility for heritability (or any other concept). It does not allow any reference to other individuals (or to parameters describing the structure of the population).[7] The intuition behind requiring irreducibility is that if heritability is to be indicative of the genetic origin of a trait, for the simple reason that the genes are properties of an individual, heritability should not also depend on the genetic composition and other properties of the population of which that individual happens to be a member.[8] Only when both criteria [at least (i') and (ii)] are satisfied will the reductionist claim that the first paragraph of this chapter notes be justified.

The conclusions of this chapter will be largely negative. It will be shown that neither of these requirements are met by either concept of heritability.[9] At least since the 1970s these conclusions have not been particularly controversial. In a recent textbook on genetic epidemiology, for instance, Khoury, Beaty, and Cohen (1993) acknowledge that heritability analysis, by itself, does not guarantee any definite genetic basis for (variation in) a trait. Rather, a high value of heritability provides only "circumstantial evidence" for genetic etiology (p. 200), indicating that further genetic analyses (using the methods that will be discussed in Chapter 5) are worthwhile. The conclusions of this chapter will suggest that even this mild claim is unduly optimistic. Note, however, that the claim made here is not simply that high heritability cannot necessarily be interpreted as a sign of a significant genetic basis of a trait. A low heritability can also not be taken to indicate that a trait, or variation in it, cannot be explained from a genetic basis.[10] However, as the discussions of Figures 4.4.1 and 4.4.2 will show, the latter is also part of the problem of interpreting heritability. These conclusions are independent of the well-known problem of the difficulty of estimating heritability accurately in all but artificially maintained experimental populations.[11] Even if heritability could be accurately estimated in practice, and there are experimental circumstances in which it can be, the interpretive problems discussed in this chapter will not disappear. Thus, in this sense, the problems with the use of heritability that are discussed here are ultimately conceptual rather than experimental.

In § 4.1 the two concepts of heritability (broad and narrow) will be explicitly defined. The theoretical basis for these definitions will be outlined in Appendix 4.1. That theory is somewhat mathematical. It is included here since it does not seem to be widely appreciated, sometimes contributing to the many misinterpretations of heritability such as the ones that are criticized in this chapter. It is relegated to an appendix so that the text itself remains accessible to those who wish to avoid mathematics. The definitions given in § 4.1 will make it immediately clear that neither criterion (i) nor (i') can be satisfied by either concept of heritability. The definitions also show that heritability is at best a relative measure of certain components of the variability of a trait (as measured by its variance) and that it is directly defined as a property of a particular population in a definite environmental range.

However, this does not show that heritability cannot somehow, indirectly, be defined as the property of an individual. Nevertheless, the failures of criteria (i) and (i′) already show that the general reductionist aim for the phenogenesis of traits cannot be achieved through heritability analysis. If what is of interest is the genesis of a trait, appealing to high values of heritability is not good enough.

However, there still remains the possibility that a high value of heritability will explain the variation in a trait from a genetic basis. In fact, geneticists have almost never claimed anything more than this, and it will be called the "minimal interpretation" of heritability in § 4.2. It will be pointed out there that even this interpretation can at best be sustained under very restrictive conditions and, once again, for a particular population in a definite environmental range. Section 4.2 will also explicitly discuss the failure of the use of heritability as a strategy for disentangling nature and nurture (to the extent that it is a well-defined problem). These negative conclusions make the pursuit of heritability analysis, which is standard fare in quantitative genetics, appear mystifying. However, narrow heritability is a useful statistic in some contexts; this will be described in § 4.3.

The analysis in §§ 4.1 and 4.2 will show that heritability is a population statistic. That it is irreducibly so, that is, there is no method by which it can even in principle be defined by reference to an individual's characteristics alone, will be shown in § 4.4. The discussion there will mainly be about broad heritability because it is that concept that has been controversially used to urge the genetic basis of the various complex human traits that were mentioned at the beginning of this chapter. However, it will also be shown that these arguments carry over to narrow heritability. These arguments will consist of the analysis of (i) a typical experimental situation and (ii) two related simple genetic models. (The mathematical analysis of these models will be relegated to Appendix 4.2.) The former will show how the restrictions necessary for the minimal interpretation of heritability are quite routinely violated. It will also show how heritability may overestimate genetic involvement. The latter will show how both concepts of heritability are irreducibly properties of populations. It will also show how heritability may underestimate genetic involvement.

Section 4.5 will note some of the problems associated with the estimation of heritability, especially in human populations. That discussion is not comprehensive; it is only included in an attempt to give some idea of the experimental problems that are also associated with the use of either concept of heritability. Finally, § 4.6 will briefly point out how broad heritability continues to be used without justification – and sometimes abused – by behavioral scientists.

Genetics and reductionism

4.1. DEFINITIONS

The basic facts about heritability that are necessary for the arguments of this chapter will be summarized in this section. Proofs of the various claims, and a systematic development of the mathematical theory on which the definitions of heritability and their interpretations are based, will be found in Appendix 4.1. Whenever claims in the rest of this chapter appear to be made without proof or argument, they should be sought in that appendix.

As already noted, the two concepts of heritability that were implicitly distinguished by Lush since about 1936, and explicitly since 1943, are narrow heritability, $\mathbf{h}$ (often symbolized as $\mathbf{h}^2$), and broad heritability, $\mathbf{H}$ (similarly, often symbolized as $\mathbf{H}^2$).[12] Their definitions are deceptively simple. Let some (phenotypic) trait be measured by a variate P, and let $\mathbf{V}$ be its variance in a given population. Let $\mathbf{V}_G$ be the part of the phenotypic variance of P (that is, $\mathbf{V}$) that can be attributed to the variation of the genotypes of the individuals of the population. Let $\mathbf{V}_g$ be the part of $\mathbf{V}$ that can be attributed to the variation in the alleles (at all the loci) in the population when they are assumed to be acting additively. Nonadditive effects at the genotypic level are the result of dominance (which, in this context, refers to any interaction between alleles at a single locus) and epistasis (which, similarly, refers to interactions between loci). Therefore, $\mathbf{V}_g \leq \mathbf{V}_G \leq \mathbf{V}$. Finally, let $\mathbf{V}_E$ be the part of $\mathbf{V}$ that can be attributed to variation in the environment. Then $\mathbf{H}$ and $\mathbf{h}$ are defined by

$$\mathbf{H} = \frac{\mathbf{V}_G}{\mathbf{V}}; \tag{4.1.1}$$

$$\mathbf{h} = \frac{\mathbf{V}_g}{\mathbf{V}}. \tag{4.1.2}$$

Therefore, $0 \leq \mathbf{h} \leq \mathbf{H} \leq 1$: narrow heritability, in this sense, is a stricter or narrower descriptive statistic than broad heritability.

Four points should be made about these definitions.

(i) Both $\mathbf{h}$ and $\mathbf{H}$ are being defined for a trait in a population in a given range of environments. They are not directly defined as properties of a trait in general, or of a trait in an individual, or even of a population in an arbitrary environment. (In particular $\mathbf{V}$ may depend on the variance of environmental variables.) This does not prove that $\mathbf{h}$ and $\mathbf{H}$ cannot, through some indirect way, be defined as a property of an individual. However, as the models analyzed in Section 4.4 will indicate by example, this option is not usually available.

(ii) These definitions do not assume that

$$\mathbf{V} = \mathbf{V}_G + \mathbf{V}_E. \tag{4.1.3}$$

Equations of this sort are generally called "additivity relations" or "linearity relations" in this context.[13] In this text it will be called the "additivity relation for **H**" or sometimes just the "additivity relation." If Equation (4.1.3) holds, then there is a potential to interpret **H** in a rather simple manner, as the fraction of the phenotypic variance that can be accounted for from variation at the genotypic level. This will be discussed in detail in § 4.2.

(iii) It is by no means obvious that the definitions of $\mathbf{V}_G$ and $\mathbf{V}_g$ – and, therefore, the definitions of **H** and **h** – as well as that of $\mathbf{V}_E$, are adequate as stated (in the sense that such a decomposition of variances is possible). The theory outlined in Appendix 4.1 provides the necessary justification.

(iv) A different point is that the definitions do not indicate how $\mathbf{V}_g$, $\mathbf{V}_G$, and $\mathbf{V}_E$ are to be measured or estimated from the data. (How **V**, which is merely the variance of P in the population, can be measured is obvious.) If the additivity relation [Equation (4.1.3)] holds, then, but only then, can $\mathbf{V}_G$ and $\mathbf{V}_E$ be directly estimated by the usual ANOVA (analysis of variance) experiments. Whether the additivity relation holds must be determined independently through experiments. Let $\mathbf{V}_d$ be that part of $\mathbf{V}_G$ that can be attributed to variation due to dominance (interpreted as the nonadditive interaction of alleles at a single locus) and $\mathbf{V}_e$ be that part that can be attributed to variation due to epistasis (all the nonadditive interactions between loci). If

$$\mathbf{V}_G = \mathbf{V}_g + \mathbf{V}_d + \mathbf{V}_e, \tag{4.1.4}$$

then ANOVA experiments can also be used to estimate $\mathbf{V}_g$. The theory developed in Appendix 4.1 shows that the status of Equation (4.1.4) is more secure than that of Equation (4.1.3) (the additivity relation) in the sense that it can be expected to hold in a greater proportion of biologically realizable situations.

(v) Because of the way in which $\mathbf{V}_G$, $\mathbf{V}_E$, $\mathbf{V}_g$, $\mathbf{V}_d$, and $\mathbf{V}_e$ have been defined, they are all measured in the same unit as **V**, that is, the unit used for the phenotypic variance. This has sometimes been regarded as problematic. The source of the worry has been that these other variances are not measured in what at an intuitive level might be called their own natural units (Bookstein 1990) or that it seems to make equations such as (4.1.3) and (4.1.4) tautological (Lewontin 1974). However, besides these intuitions, no explicit argument has been given regarding why the use of units inherited from the phenotypic variance would lead to any misuse of either the other variances or of heritability. This point will not be considered as an objection to the use of heritability any further in this chapter.

4.2. INTERPRETATION: NATURE, NURTURE, AND REDUCTION

Among geneticists, the proponents of the use of heritability have traditionally regarded **H** as the relative (that is, fractional) measure of the extent to which the variability of a trait (as measured by the variance) can be attributed to variability at the genotypic (or, as it is usually put, genetic) level.[14] This will be called the minimal interpretation of **H**. It provides the basis for the more ambitious interpretations that have sometimes been used. Similarly, the minimal interpretation of **h** will be to regard it as the relative measure of the extent to which the variability of a trait can be attributed to the variability of the alleles (at all loci) when they are presumed to be acting additively (that is, completely independently of each other).

The theory developed in Appendix A.1 shows that these interpretations are tenable if and only if the additivity relation [Equation (4.1.3)] holds, provided that **h** and **H** are taken to be properties of the population for the particular environments for which the different variances are measured.[15] Further, the additivity relation holds if and only if the genotypes and environments are statistically uncorrelated throughout the range being considered (their covariance is equal to 0) and provided that their interaction is a constant (its variance is equal to 0). Moreover, in general, nothing can be said about individuals or about the population at a different time (unless all genotypic and allelic frequencies remain constant and the environmental variables remain within the original range): when these conditions are met,

$$\mathbf{H} = \frac{\mathbf{V}_G}{\mathbf{V}_G + \mathbf{V}_E}; \tag{4.2.1}$$

and

$$\mathbf{h} = \frac{\mathbf{V}_g}{\mathbf{V}_G + \mathbf{V}_E}. \tag{4.2.2}$$

If the conditions mentioned above do not hold, there are additional terms in the denominator. One consequence of this is that $1 - \mathbf{H}$ cannot be regarded as the environmental component of the variance, even if one continues to regard **H** as the genotypic component. It is possible to tolerate this situation by choosing never to refer to $1 - \mathbf{H}$. But even such a move proves inadequate.

The additional terms in the denominator routinely depend on V_G. Consequently, H is no longer a linear function of V_G and, therefore, cannot be a measure of the influence of the genotypic component on V. A high value of H compared with a low one no longer necessarily represents a proportional increase in the relative contribution of genotypic variance to the phenotypic variance.

Moreover, one of the additional terms due to genotype–environment correlations may be a covariance. It can be positive or negative if the genotypes and environments are correlated (depending on whether the correlation is positive or negative). However, the other term representing a genotype–environment interaction is a variance and, therefore, always positive. This leaves open the possibility that H can at least be regarded as a lower limit for the contribution of the genotypic variance to the phenotypic variance even when there is an interaction or only a positive correlation between genotypes and environments. However, a move of this sort does not answer the point about nonlinearity that was made above. Moreover, the example and the models analyzed in § 4.4 will show its inadequacy in another way. All these objections carry over to h: Sober and Wilson's (1994) claim that "heredity" can be measured by h is an error.

What is perhaps worse is that, even if additivity and, therefore, Equations (4.2.1) and (4.2.2) hold, a further caveat is necessary. Restricting attention to H, it is at best an estimate of the relative importance of genetic factors in the variability of the trait and not on the value that it has (even on the average). The reason for this is quite simple: two statistical distributions may have the same variance but different means, and H only reflects the ratios of the variances.[16] The same caveat, now referring to the "effects of alleles acting additively," rather than "genetic factors" in general, is also obviously necessary for h. Thus, the reductionist claim with which this chapter began has no chance of being even approximately true (see below).

Going beyond the partly tenable minimal interpretations, two additional claims about H have routinely been made. These will be considered next. Both of them suffer from all problems of the minimal interpretation, including those that have already been mentioned and others that will be discussed in § 4.4. Moreover, these additional claims also involve new contentious assumptions.

(i) A long-standing view of H has been that it helps sort out the relative importance of "nature" or "nurture" in the "nature–nurture" dispute.[17] Even after identifying nature with biological nature – and there may well be reasons to question that – this interpretation involves a further conflation of the categories of the biological and the genetic (see also Chapter

7, § 7.1). In defense of this conflation, it can be pointed out that if nature is to be contrasted with nurture as is implicitly implied by referring to the "nature–nurture dispute" as a "dispute," then nature should be construed not as all biological factors but only the inherited ones. Nevertheless, it is at least equally reasonable to suggest that the nature–nurture distinction is one between the relative role of biological (as recognized by structural properties of biological organisms) and nonbiological factors in the explanation of some phenomenon.[18] Moreover, more than genes are inherited, and not everything genetic is (see Chapter 7, § 7.1).

Even if "nature" is "biological nature," not all biological factors are necessarily genetic. Physical or mental retardation in humans, when induced by malnutrition, is obviously biological. It is not genetic; it can occur irrespective of the genotype of an individual. To use a more controversial example, even if it is true that there is some biological mechanism by which different morphological properties of brains lead to different sexual preferences or behaviors [as originally claimed by LeVay (1993)], it is still possible that these properties, and therefore these preferences or behaviors, are not to the same extent genetic. Non-genetic factors may have influenced brain morphology, especially during development. Therefore, using **H** to attempt to disentangle nature from nurture not only suffers from all the interpretive problems of **H** in its minimal interpretation but involves an additional conflation of some importance. Strictly speaking, the "nature–nurture" dispute, including the question of whether it is even well-defined, let alone how it is to be resolved, is beyond the scope of this book. Only a few peripheral remarks about it will be made below.

(ii) It has routinely been implicitly suggested – even by those such as Plomin (e.g., 1994) who explicitly emphasize that **H** depends on the composition of a population – that **H** is the property of a trait itself and not just a trait in a given population (at a particular time). This suggestion is implicit every time there is mention of the heritability of a trait without explicit reference to a particular population. Going slightly further, it is some-times at least implicitly suggested that **H** quantitatively estimates at least the extent to which a trait can be attributed to genetic factors. Slightly more modestly, and much more plausibly, it is routinely suggested that **H** quantitatively estimates the extent to which the variation in a trait among individuals can be attributed to these factors.[19] The definitions show that all such claims are misleading at best: **H** (or **h**) has no meaning unless relativized to a particular population and to the set of relevant environments.

Returning to the question of reduction, the fundamentalist assumption in putative reductions using heritability analysis (by invoking a high value for **H**) is simply that the genetic level (including the mechanisms and the rules that are operative there) is more fundamental than the phenotypic one. Note, moreover, in the use of **H**, that no other property – not even any abstract structural property – is attributed to the genotype.[20] Returning to the classification of the last chapter (§ 3.2), this would be a weak reduction [that is, of type (a)]. The most interesting question is the status of such reductions, that is, the extent to which they are successful. As has already been noted at the beginning of this chapter, the question of whether this interpretation of **H** is justified depends on whether **H** satisfies the criteria (ii) and at least (i') [preferably (i)] that were listed there. It should be emphasized that these criteria go well beyond the minimal interpretation. Therefore, given that the definitions already show that (i') is violated, there is no potential for such a reduction. But suppose that one gives up the original (and interesting) reductionist aim and simply attempts to reduce the variability of the trait to the genotypic realm. That is simply the question of the validity of the minimal interpretation; § 4.4 will show its limitations. (Alternatively, Appendix 4.1 establishes the exact conditions under which it is tenable.)

4.3.   THE USE OF **h**

The interpretive problems noted in the previous two sections (which arise from the mathematical analysis of Appendix 4.1) show that **H** and **h** can only have their minimal interpretations in rather special circumstances. This naturally raises the question whether they can play any significant conceptual or theoretical role in genetics. However, even if they cannot, it is still possible that they may have some pragmatic value that would justify the continued attempts at their estimation. For **h** there is, indeed, such a value.

Roughly, the effectiveness of selection, measured by its rate, is proportional to **h**, a result of significance to animal and plant breeders. Suppose that environments and genotypes are uncorrelated. (This can be ensured by intervention in an experimentally manipulated population.) Further, suppose that the additivity relation [Equation (4.1.3)] holds, at least approximately. In other words, it will be assumed that there is no genotype–environment correlation (from a covariance or an interaction). Let $\bar{P}_n$ be the average value of some (continuous) variate of a phenotypic trait in the n-th generation, $\bar{P}_{n+1}$ its average value in the $(n + 1)$-th generation. Let $\bar{P}_n^s$ be its average value in the n-th generation after selection according to some criterion. Let

$$\Delta \bar{P}_n = \bar{P}_{n+1} - \bar{P}_n. \tag{4.3.1}$$

Let

$$S_n = \bar{P}_n^s - \bar{P}_n. \tag{4.3.2}$$

Let **h** be the narrow heritability of the trait in question in that population and for the environments that have been considered. Then, under a wide variety of types of selection:

$$\Delta \bar{P}_n = \mathbf{h} S_n. \tag{4.3.3}$$

(For a proof of this assertion and a careful treatment of the types of selection and assumptions about the breeding patterns of populations for which Equation (4.3.1) holds, see Nagylaki (1992, pp. 311–20).) This is an analog of what, in the context of models of selection with discontinuous traits, Fisher (1930) called the "fundamental theorem of natural selection." In both cases, there are many situations when the assumptions required for these derivations are not met. However, that problem is marginal to those being discussed in this book and will not be pursued any further here.

It should be clear why Equation (4.3.3) is valuable to a breeder who can manipulate a population enough to ensure that there is no correlation between the genotypes and the environment of a population. If **h** is small for a particular trait in a population, selection for individuals with a desired value for that trait will not be effective. In the case of **H**, there is no similar general result. If **H** has any scientific value, it would have to be at the conceptual or theoretical level. The interpretive problems noted in the previous section already give ample reason for skepticism on this point – but there is worse to come.

4.4. CONCEPTUAL PROBLEMS ASSOCIATED WITH **H**

Returning to the question of the range of validity and implications of the minimal interpretation, it will now be shown exactly how the failure of the additivity relation [Equation (4.1.3)] leads to serious problems for the interpretation of **H**, and also that even when the minimal interpretations are tenable, **H** (or **h**) cannot still be used for genetic explanation. To do this, the following two cases will be analyzed separately and explicitly: (i) when the additivity relation [Equation (4.1.3)] fails; and (ii) when it holds and, therefore, the minimal interpretation is tenable.

(i) The basic feature of this case is that there is a statistical correlation between G and E, no matter what the source of that correlation is. There are two ways in which such a correlation may be generated: (a) there is a statistical correlation between the genotypes and environments, and (b) there is a varying interaction between the genotype and environment.[21] While the argument that will be given here is applicable to either of these two situations, only the latter will be explicitly treated for two reasons:[22] (a) for many traits, whether there is a significant interaction between a genotype and its environment is often a biological property of a population (and sometimes even of a species) that cannot be manipulated by an experimentalist (or breeder).[23] Thus, even **h** may be of little value for the purpose of selection. And (b), as was briefly noted in § 4.2, the presence of a varying interaction still leaves open the move of salvaging **H** (or **h**) by arguing that it provides a lower limit for the actual contribution of genetic factors to the phenotypic variance [when it is measured in a population and interpreted according to Equation (4.3.14)]. The argument given here will respond to that move.

The rudiments of this argument go back to Hogben (1933) who argued, prophetically, that it showed the "interdependence" of nature and nurture.[24] The details of the argument given here are mostly due to Layzer (1974). In Figure 4.4.1, the phenotypic value (P) of a continuous

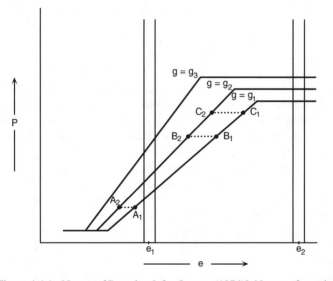

Figure 4.4.1. Norms of Reaction [after Layzer (1974)]. Norms of reaction for three genotypes in a hypothetical population. *Abcissa*: the value of an environmental variable. *Ordinate*: phenotypic value. For a discussion, see text.

trait is plotted against an environmental variable for three different genotypes in a hypothetical example. Following Woltereck (1909), a curve
of this sort is usually called a "norm of reaction." Consider three pairs
of points (A_1, A_2), (B_1, B_2), and (C_1, C_2). For each pair, $\Delta P = 0$, and
ΔG is constant. Therefore, if the interaction is to be equal to 0, then ΔE
must be the same for all three pairs, which is not the case. Moreover,
the difference in ΔE between (A_1, A_2) and (B_1, B_2) is not equal to that
between (A_1, A_2) and (C_1, C_2); consequently, the interaction cannot
even be a constant. Therefore, the additivity relation [Equation (4.1.3)]
will fail.

Of course, the important question is whether – and if so how often –
norms of reaction are encountered in nature. Hogben (1933) provided an
example as early as 1933 drawing on Krafka's (1920) data (see Figure
4.4.2). In Hogben's example, P is the number of facets in the compound
eye of Drosophila, E is the temperature. Lewontin (1974) points out that
such a situation is standard (see Figure 4.4.3), and also discusses the various forms the norms of reaction may take. Only if all norms of reaction
for the different genotypes of a population are parallel to each other can
this problem be avoided.

In Figure 4.4.1, consider two narrow ranges of values of E around e_1
and e_2. In the range around e_1, **h** and **H** will both be almost equal to 1.
Yet, as Lewontin (1970) has pointed out, the phenotypic variability in
this region should be attributed to environmental factors given that, for
the same amount of genotypic variability, there is so much more phenotypic variability at e_1 than in a range of environments (of the same size)
around e_2. Since in almost all circumstances **H** and **h** have to be estimated
from data from a narrow range of environmental values, this example
shows that their values cannot be reliably used to apportion explanatory
weight between genetic and environmental factors, even when all that
is sought is an explanation for phenotypic variability. Now, choose any
environment at which the norms of reaction are relatively horizontal.
Consider an environment around the chosen point. By decreasing the
environmental range, $\mathbf{V}(E)$ can be made to decrease faster than $\mathbf{V}$.[25]
Consequently, **H** is driven upward with no change taking place in the
genetic composition of the population or the role of genes in the origin
of the trait. In this sense, **H** progressively overestimates the extent to
which the trait can be accounted for by genetic factors. Finally, **H** and **h**
do not permit any understanding of the interesting details of the various
norms of reaction: their differing slopes, thresholds, and final values
(Layzer 1974).

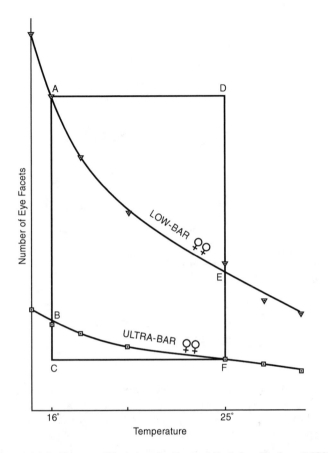

Figure 4.4.2. Norms of Reaction for Drosophila [after Hogben (1933), p. 384]. The graphs, as indicated, are for the *low-Bar* and the *ultra-Bar* genotypes. *Abscissa*: temperature; *Ordinate*: number of eye facets. The data are from Kafka (1920). The important point is that the norms of reaction are not parallel to each other.

(ii) What if perfect additivity (Equation 4.1.3) holds? As mentioned before, in such a circumstance, the minimal interpretations of both **h** and **H** are tenable. But is it possible to go any further and adopt either of the two more ambitious interpretations discussed in § 4.2? The answer is "no" because **h** and **H** are both functions of the frequency of the different genotypes in the population and also depend on other population

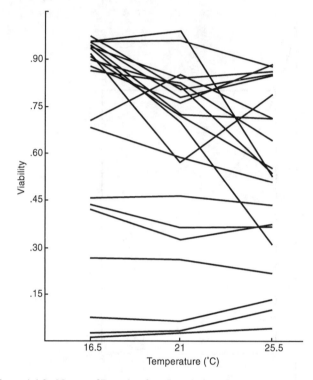

Figure 4.4.3. Norms of Reaction for a Population of *Drosophila pseudoobscure* [after Lewontin (1974); by permission of the University of Chicago Press]. Each norm corresponds to a different genotype. *Abscissa*: temperature; *Ordinate*: viability. The data are from Dobzhansky and Spassky (1944). The huge variation in norms of reaction seen here is probably typical of all large populations.

characteristics such as mating patterns. The exact nature of these functions depends on environmental and population parameters but, except possibly in very contrived cases, this frequency–dependence is unavoidable. Thus, **h** and **H** are irreducibly population characteristics – they cannot be defined using the properties of an individual alone.

This point is best illustrated by explicitly analyzing simple population genetic models in which the exact genetic and environmental bases of some phenotypic trait will be assumed to be known.[26] Of course, in practice, if these are known, there would be no reason for estimating **H**. (**h** could still be useful since it would indicate how effective, in terms of speed of phenotypic change, selection would be.) However, the point being pursued here is the conceptual clarification of these parameters. For that purpose, explicit genetic models such as the ones that will now be developed are helpful. However, note that the models treated here assume a single locus and are, therefore, of very limited direct applicability. Certainly, they are not applicable to quantitative traits that depend on a large number of loci and are usually the ones for which heritability analysis is attempted. Nevertheless, past experience in population genetics, especially in the various proofs of the so-called fundamental theorem or natural selection (which involve analyses of genetic variance), suggests that the conclusions reached with a single-locus model about the frequency-dependence of variances will be generally true of multiple-locus models. It should be emphasized, however, that the present status of this claim is only that of a somewhat educated conjecture.

h will be treated first. The model analyzed here is due to Kempthorne (1969, pp. 308–15); the treatment is partly due to Jacquard (1983). The model assumes that some phenotypic trait is entirely under the control of one locus with two alleles, A and a; that is, the value of that trait for an individual depends entirely on that individual's genotype. In other words, environmental factors have no role in accounting for any observed variation in the trait. (Therefore, **H** will be equal to 1.) It will be assumed that mating is random with respect to this locus, and that the population is infinite. It will turn out that **h** will be a rather awkward (in the sense of being a highly variable) function of allele frequencies even in such a model.

This model is described by Table 4.4.1. Let p be the frequency of A and q be the frequency of a ($p + q = 1$). Because mating is at random with respect to this locus, the genotypic frequencies of AA, Aa, and aa are p^2, $2pq$, and q^2, respectively. The table gives the genotypic values, that is, the part of the phenotypic value contributed by the genotype (in this case, everything) for each of the genotypes and the "coded genotypic value," which is defined as the difference between the genotypic value (of each genotype) and the mean value for the population. Thus, d is the genotypic value of AA; h is the genotypic value of Aa; and r is that for aa. μ is the mean value. Then, the coded genotpic values for AA, Aa, and aa, which are i, j, and k (respectively), are simply equal to $d - \mu$, $h - \mu$ and $r - \mu$. Such coded values or differences from the mean are the customary variables used in quantitative genetics (following a tradition that goes back to biometry).

87

Table 4.4.1.   *Model of a genetic trait controlled at one locus*

Genotype	Frequency	Genotypic value	Coded genotypic value
AA	p^2	d	$i = d - \mu$
Aa	$2pq$	h	$j = h - \mu$
aa	q^2	r	$k = r - \mu$

Note: There are two alleles A and a controlling the trait; p is the frequency of A, $q = 1 - p$ is the frequency of a. It is assumed that mating is at random (with respect to this locus). Therefore, the frequencies of the genotypes, AA, Aa, and aa, are obtained by multiplying the frequencies of the individual alleles. The phenotypic values of the trait displayed by these three genotypes are d, h, and r, respectively. Since the model assumes no environmental influence, the genotypic values are equal to the phenotypic values. μ is the mean genotypic value of the trait in the population. The coded genotypic value, for each genotype, is the difference of its genotypic value and the mean for the population.

It is shown in Appendix 4.2 that the narrow heritability **h** is given by

$$\mathbf{h} = \frac{[p(i - j) + q(j - k)]^2}{[p(i - j) + q(j - k)]^2 + pq[(i - j) - (j - k)]^2}. \qquad (4.4.1)$$

Define $m = i - j$ and $n = j - k$. Then m and n may be taken to be independent and **h** can be written in the form

$$\mathbf{h} = \frac{[pm + (1 - p)n]^2}{[pm + (1 - p)n]^2 + p(1 - p)(m - n)^2}. \qquad (4.4.2)$$

This form for **h** is particularly easy to analyze. It is shown in Appendix A.2 that **h** can take potentially any value between 0 and 1 provided that the genotypic value for the heterozygote is either above or below that of both homozygotes. In general, **h** becomes close to 1 when one or the other of the alleles is near fixation. Figure 4.4.4 shows this dependence of **h** on p.

   To see how **H** can be frequency-dependent, consider a second model that differs from the first only insofar as there is an environmental dependence of the phenotype, but no genotype–environment correlation (through either covariance or interaction). Thus, the minimal interpretation is still tenable. For simplicity, assume that $\mathbf{V}_E$ is a constant.

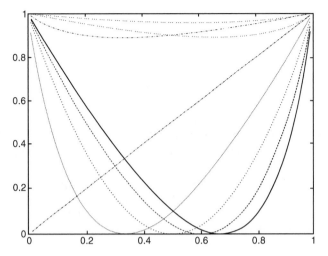

Figure 4.4.4. Frequency-Dependence of the Narrow Heritability of a Genetically Controlled Trait. Narrow heritability, **h**, as a function of the frequency, p, of allele A. For a discussion of the model and an explanation of the parameters, see the text. $m = 1$: $n = -2$: ———; $n = -1.5$: – – –; $n = -1.0$: · · · · · ·; $n = -0.5$: ·····; $n = 0.0$: —··—; $n = 0.5$: -----; $n = 1.0$: ···· ···· ····; $n = 1.5$: ··· ··· ···; $n = 2.0$: ·· ·· ··.

It is shown in Appendix 4.2 that

$$\mathbf{H} = \frac{[pm + (1 - p)n]^2 + p(1 - p)(m - n)^2}{[pm + (1 - p)n]^2 + p(1 - p)(m - n)^2 + \mathbf{V}_E}. \tag{4.4.3}$$

H is obviously frequency-dependent. What is more intriguing is that it is a monotonically increasing function of p. Thus, a high **H** may only indicate a high frequency of the A-allele.

It should be obvious that neither criterion (i′) [let alone (i)] nor criterion (ii) from the beginning of the chapter is satisfied by either **h** or **H** in either of these models.[27] Claims that a high value for either provides an explanation for a phenotypic trait, or even differences in values of the trait, are simply misguided. Note that (i) this conclusion is primarily the result of the conceptual analyses attempted in this section, especially the analysis of the model described by Table 4.4.1; (ii) however, to the extent that the norms of reaction in Figure 4.4.1 are also being used to urge the implausibility of using high values of **h** or **H** to infer a genetic explanation of phenotypic differences, there is an empirical assumption involved. That assumption is that norms of reaction are not parallel to each other. This is not a particularly strong assumption. Nevertheless, at least in that part of the argument of this section, empirical considerations are interwoven with conceptual ones to argue for the conclusion reached here.

Thus, heritability analysis cannot be used even to reduce phenotypic differences, let alone phenotypic values, to a genetic basis. Though it is not a major concern of this book, the failure of criteria (i'), (i), and (ii) can also be used to argue that heritability analysis cannot be used to disentangle nature from nurture. A high value of **H** (or **h**) does not necessarily indicate a strong influence of natural factors (where those are being interpreted as genetical ones). Conversely, a low value of **H** (or **h**) does not indicate a strong influence of environmental ones. And if this is not counterintuitive enough, the values depend on the composition of the population.

4.5.   PROBLEMS OF ESTIMATION

In general, experimental issues are beyond the scope of this book. Nevertheless, in both the cases of **h** and **H**, to complete the discussion of the problems associated with their customary use a few cautionary notes about experimental topics are in order. In the case of experimental populations whose environments and genotypic composition can be systematically manipulated, the additivity relations [Equations (4.1.3) and (4.1.4)] can be made to hold. ANOVA experiments – where a genotypic variable is held constant while the environmental variables are systematically varied, and vice versa – can then be used to estimate the different components of the phenotypic variance, and **H** and **h** can be calculated (for that population in that range of environments).

In human populations such a procedure is infeasible for obvious ethical reasons. Those who would measure heritability usually restrict attention to **H** (which, in general, is the easier parameter to estimate).[28] However, unless the additivity relation [Equation (4.1.3)] holds, the standard techniques for estimating **H**, such as twin or adoption studies, are invalid.[29] Recall that Equation (4.1.3) is also necessary for the minimal interpretation of **H** to be tenable. It is only under that interpretation that **H** is the fraction of the phenotypic variance that can potentially be explained by variation at the genetic level. Thus, the interpretive desire to have a meaningful **H** and the epistemological desire to be able to estimate it lead to the same requirement, the additivity assumption.

There is no a priori reason to assume the additivity relation [Equation (4.1.3)]: rather, given the extent to which humans manipulate their environments, a priori considerations would suggest that there would always be significant correlations between genotypic and environmental variables. Those who advocate the estimation of heritability using the standard techniques resort to empirical considerations. Plomin (e.g., 1994) and others argue that the covariance and interaction terms in Equation (4.3.6) have been found to be small from all experiments on human populations. This would mean that **H** can be at least approximately estimated in practice, and the minimal interpretation used for what would amount to a rough explanation of phenotypic differences from genotypic ones. The trouble, however, is that the standard tests used to judge the presence of genotype–environment covariance or interaction are known to be insensitive. Wahlsten (1990) has shown that these tests would routinely not detect even the nonlinearity of the (Newtonian) gravitational interaction. Moreover, norms of reaction such as those graphed in Figures 4.4.1, 4.4.2, and 4.4.3 are endemic in animal populations. No such graphs are apparently available for human populations, but it is hard to believe that those norms or reaction would always be parallel to each other. The experimental situation looks no more promising than the interpretive one for advocates of the use of **H**, especially for human populations. Finally, and perhaps most importantly, even if **H** can have its minimal interpretation, the fact that it does not lead to any genetic explanation, as shown in § 4.4, makes it rather devoid of intellectual interest.

4.6. THE ABUSE OF **H**

The utility of **h** to the animal or plant breeder was indicated in § 4.4. However, for **H** there is no result analogous to Equation (4.3.3) in generality or usefulness. Moreover, as the discussion in § 4.5 showed, neither **h** nor **H** [even when the minimal interpretation is tenable, that is, the additivity relations of Equations (4.1.3) and (4.1.4) hold] can play any cogent explanatory role if the question is even one of explaining the origin of phenotypic traits (and not just their temporary variability) in a population, let alone that of explaining the origin of these traits in an individual. So, why bother with **H**? There is no good reason. Bouchard et al. (1990), for example, argue that an alleged high value of **H** for IQ shows that general intelligence is strongly affected by genetic factors. But, as the discussion of § 4.5 shows, this is a fallacy.

Crow (1990) has argued that **H** is interesting and important by itself and, moreover, provided that Equation (4.1.3) holds, values of **H** can be used to judge whether environmental intervention can change the distribution of a trait:

> One can be interested in the heritability of IQ without advocating a breeding program. A [**H**] of 0.2, based on identical genotypes in independent environments, tells us that 80 percent of the variance is environmental. This says that changes of existing variables within the existing range can have a substantial effect on the trait. On the other hand if the [**H**] is high, environmental manipulations will have little effect unless they extend beyond the existing range, or bring in new factors." (p. 127)

Crow is using the minimal interpretation, and his conclusions are correct so long as the additivity relations necessary for the validity of that interpretation are met. Once again, as the discussion in § 4.4 showed, **H** can be high (or low) simply because of genotype–environment covariances or interactions, choice of environmental range, or its dependence on allelic frequencies, without necessarily indicating anything about the respective roles of the genotype and environment. The most important point that Crow makes is one that he only implicitly alludes to: even if **H** is high and the minimal interpretation holds, changing the environmental variables beyond the current range can significantly alter the distribution of a trait (including its average value).

In an odd move, Plomin has argued that research into **H** is the best way to demonstrate the importance of environmental factors on a trait.[30] Presumably, a low value for **H** is to be interpreted as a sign of environmental influence – this is invalid (as, once again, the discussion in § 4.4) shows. However, if additivity [Equation (4.1.3)] is assumed, and estimates of **H** using different strategies disagree, then genotype–environment correlations of different types are suggested.[31] The trouble is that the detection of an environmental influence or even a genotype–environment interaction through heritability analysis gives no idea of what that influence or interaction might be: there are no simple rules of environmental transmission or genotype–environment interaction (Crow 1990). To go any further requires the construction of explicit models, such as the one analyzed in § 4.4, that is, standard Mendelian or molecular genetics, which will be discussed in the next two chapters. Crusio (1990) and others seem to have assessed the situation correctly when they have argued that, at the very least, the time for heritability analysis is long past.

It is hard, therefore, not to suspect that the continued pursuit of **H** is guided, at least to some extent, by noncognitive, especially, political factors.

There are at least three sets of considerations that point to such a conclusion.

(i)   Much of the analysis of **H** has concentrated on human traits, especially psychological traits, and has been pursued much more often by social scientists rather than bona fide geneticists. Meanwhile, human geneticists, starting with the pioneering work of Hogben and Haldane in the 1930s, have largely pursued conventional analyses such as segregation and linkage analyses.[32]

(ii) The traits for which **H** continues to be pursued often include those carrying social judgments, even if they are ill-defined. For instance, Bouchard et al. (1990) report high values of **H** for something called "religiosity" as well as IQ.[33] Bouchard (1994) reports relatively high values of **H** for "openness," "agreeableness," "conscientiousness," "neuroticism," and "extroversion," each of which is a trait that carries social judgment.

(iii) Meanwhile the work on **H** and IQ has been routinely used to argue for genetic inferiority of certain ethnic groups, particularly African-Americans. This argument flourished in the early 1970s and has been revived again in the 1990s.[34] It is based on the assumption that if IQ has a high value for **H** in two different populations, then it can be concluded that the difference in mean IQ between the two populations also has a high group heritability (which, incidentally, cannot even be defined in the way **h** or **H** is defined), an assumption that is usually admitted as having no known factual basis even by those who advocate the use of **H**.[35]

Perhaps the increase in sophistication of genetical techniques, and their incorporation into behavioral studies, partly a result of the Human Genome Project (HGP), will finally shift attention away from **H**. If so, the HGP will have served at least some useful purpose. Meanwhile, if political considerations are left aside, what the discussions in this chapter show is that it is at least very hard, if not impossible, to account for some of the behavioral scientists' continued obsession with heritability.

APPENDIX 4.1

Theory

The theoretical basis for the analysis of heritability has been particularly clearly outlined by Layzer (1974), and the exposition here will generally follow that treatment.[36] Let $P(\vec{g}, \vec{e})$ be the value of some phenotypic character, that is, a phenotypic value, where $\vec{g}$ is an n-dimensional vector of genotypic variables which, together, specify a genotype; and $\vec{e}$ is an m-dimensional vector of environmental variables similarly specifying an environment. The components of $\vec{g}$ and $\vec{e}$ can each be regarded as random variables. Let the joint probability distribution of all these random variables be specified by $\Phi(\vec{g}, \vec{e})$. Let the joint probability distribution of the elements of $\vec{g}$ be specified by $\Phi_1(\vec{g})$ and that of the elements of $\vec{e}$ be specified by $\Phi_2(\vec{e})$, that is,

$$\Phi_1(\vec{g}) = \int_1 \cdots \int_n \Phi(\vec{g}, \vec{e}) \, d\vec{e}; \qquad (4.A1.1a)$$

$$\Phi_2(\vec{e}) = \int_2 \cdots \int_m \Phi(\vec{g}, \vec{e}) \, d\vec{g}. \qquad (4.A1.1b)$$

The conditional probability distributions $\Phi(\vec{g} \mid \vec{e})$ and $\Phi(\vec{e} \mid \vec{g})$ are then defined by

$$\Phi(\vec{g} \mid \vec{e}) = \frac{\Phi(\vec{g}, \vec{e})}{\Phi_2(\vec{e})}; \qquad (4.A1.2a)$$

$$\Phi(\vec{e} \mid \vec{g}) = \frac{\Phi(\vec{g}, \vec{e})}{\Phi_1(\vec{g})}. \qquad (4.A1.2b)$$

Following Fisher (1918) and Layzer (1974), the genotypic value of a genotype, $G(\vec{g})$, that is, the average contribution to the phenotypic value made by that genotype across the range of environments being considered, can now be expressed as

$$G(\vec{g}) = \int_1 \cdots \int_m P(\vec{g}, \vec{e}) \Phi(\vec{g} \mid \vec{e}) \, d\vec{e} \qquad (4.A1.3)$$

Similarly, the environmental value, $E(\vec{e})$, of an environment is defined by

$$E(\vec{e}) = \int_1 \cdots \int_n P(\vec{g}, \vec{e}) \Phi(\vec{e} \mid \vec{g}) \, d\vec{g} \qquad (4.A1.4)$$

This is the average contribution of that environment to the phenotypic value across the range of genotypes being considered.

A new (remainder) function, $R(\vec{g}, \vec{e})$, is now defined by

$$R(\vec{g}, \vec{e}) = P(\vec{g}, \vec{e}) - G(\vec{g}) - E(\vec{e}). \qquad (4.A1.5)$$

$R(\vec{g}, \vec{e})$ can be interpreted as the part of the phenotypic value that is not accounted for by a sum of the genotypic and environmental values. From

Equation (4.A1.5),

$$\mathbf{V}(P) = \mathbf{V}(G) + \mathbf{V}(E) + 2\mathbf{Cv}(G, E) + \mathbf{I}(G, E, R) \qquad (4.\text{A}1.6)$$

where $\mathbf{V}(\ )$ and $\mathbf{Cv}(\ ,\ )$ denote the variance and covariance, respectively, and

$$\mathbf{I}(G, E, R) = 2\mathbf{Cv}(G, R) + 2\mathbf{Cv}(E, R) + \mathbf{V}(R) \qquad (4.\text{A}1.7)$$

is the contribution of the genotype–environment interaction to the phenotypic variance.[37]

Now the covariance of G and E, $\mathbf{Cv}(G, E)$, vanishes if and only if the genotypic and environmental variables are statistically independent, that is

$$\Phi(\vec{g}, \vec{e}) = \Phi_1(\vec{g})\Phi_2(\vec{e}). \qquad (4.\text{A}1.8)$$

Moreover, if Equation (4.A1.8) holds,

$$\mathbf{Cv}(G, R) = \mathbf{Cv}(E, R) = 0. \qquad (4.\text{A}1.9)$$

[see Layzer (1974) for a proof];[38] that is,

$$\mathbf{I}(G, E, R) = \mathbf{V}(R).^{[39]} \qquad (4.\text{A}1.10)$$

The broad heritability, $\mathbf{H}$, can now be specified by

$$\mathbf{H} = \frac{\mathbf{V}(G)}{\mathbf{V}(P)}$$

$$= \frac{\mathbf{V}(G)}{\mathbf{V}(G) + \mathbf{V}(E) + 2\mathbf{Cv}(G, E) + 2\mathbf{Cv}(G, R) + 2\mathbf{Cv}(E, R) + \mathbf{V}(R)} \qquad (4.\text{A}1.11)$$

If Equation (4.A1.8) holds, this simplifies to

$$\mathbf{H} = \frac{\mathbf{V}(G)}{\mathbf{V}(G) + \mathbf{V}(E) + \mathbf{V}(R)}. \qquad (4.\text{A}1.12)$$

Note that Equation (4.A1.8) is not sufficient to get the additivity relation, that is, Equation (4.1.3). To obtain that simple relation, the condition

$$\mathbf{V}(R) = 0 \qquad (4.\text{A}1.13)$$

must be independently imposed. This means that not only is there no G-E covariance, but also that the remainder left in the phenotypic value after the genotypic and environmental components are considered is a constant; it is tantamount to assuming that $R(\vec{g}, \vec{e}) = 0$.[40] Finally, if both Equations (4.A1.8) and (4.A1.13) hold:

$$\mathbf{H} = \frac{\mathbf{V}(G)}{\mathbf{V}(G) + \mathbf{V}(E)}. \tag{4.A1.14}$$

In Equation (4.A1.14), **H** can plausibly be interpreted as the fraction of the total phenotypic variance contributed by the genotypic variance – in other words, the minimal interpretation is tenable. It is instructive to analyze exactly what happens when either Equation (4.A1.8) or Equation (4.A1.13) is not true. In either case, since the denominator would contain a term other than $\mathbf{V}(G)$, which also depends on G: (i) $1 - \mathbf{H}$ would not measure the environmental variability's contribution to the phenotypic variance; and (ii) **H** would no longer (necessarily) be a linear function of $\mathbf{V}(G)$. While (i) may be somewhat irritating, it is not a serious objection to the interpretation of **H**. One would simply not interpret $1 - \mathbf{H}$ as the environmental component. However, (ii) – the loss of linearity – suffices to rule out the minimal interpretation: a high value of **H** compared with a low one would not necessarily represent a proportional increase in the relative contribution of the variability of genetic factors to the phenotypic variance.

Beyond this, there is an important asymmetry between the effects of failures of Equations (4.A1.8) and (4.A1.13). Suppose that Equation (4.A1.8) holds, while Equation (4.A1.13) does not. Since a variance is always a positive quantity, the presence of the nonzero $\mathbf{V}(R)$ in the denominator in Equation (4.A1.12) would make an experimentally evaluated **H** underestimate the genetic, as opposed to the environmental contribution to $\mathbf{V}(P)$, if that **H** is interpreted according to Equation (4.A1.14). One could still potentially argue that a high value for **H** indicates a high contribution to $\mathbf{V}(P)$ from the genetic level. The additional interpretive problems that make this move insufficient are discussed in § 4.4 (see the discussion of Figure 4.4.1). In contrast to this situation, since covariances can potentially be negative, if Equation (4.A1.8) fails, the opposite can happen irrespective of whether Equation (4.A1.13) holds. Therefore, unless Equation (4.A1.8) holds, **H** should not be used as a measure of genotypic influence on the variation of a trait even for pragmatic purposes.[41]

That these interpretive problems carry over to the definition of narrow heritability, **h**, should be obvious. What is less obvious is that no additional interpretive problems arise, provided that there is no linkage. To define **h**, $G(\vec{g})$ has to be decomposed further. Assume that the population is infinite, random mating, and diploid, and that there is no linkage. Then $G(\vec{g})$ can be written in the form

$$G(\vec{g}) = G_a(\vec{g}) + G_d(\vec{g}) + G_e(\vec{g}) \qquad (4.\text{A}1.15)$$

where $G_a(\vec{g})$ is the sum of the contributions of the alleles acting independently (the "additive" component), $G_d(\vec{g})$ is the sum of the residual contributions of the two alleles at each locus (the "dominance" component), and $G_e(\vec{g})$ is the sum of the still remaining contributions of the loci taken together (the "epistatic" component). The statistical independence assumptions incorporated in Mendel's laws of segregation of alleles (at each locus) and independent assortment (of loci) ensure that (Layzer 1974)

$$\mathbf{V}(G) = \mathbf{V}(G_a) + \mathbf{V}(G_d) + \mathbf{V}(G_e). \qquad (4.\text{A}1.16)$$

No new covariance terms appear. Consequently, the minimal interpretation of **h** is tenable if and only if Equations (4.A1.8) and (4.A1.13) hold. Layzer's argument is not obviously extendible to the situation where there is linkage; a detailed analysis of that problem apparently has never been performed.

Finally, the way in which the genotypic and environmental values have been defined show that nothing in general can be said about **H** or **h**, if the genotypic or environmental values go beyond the ranges of integration in Equations (4.A1.1a, b). If those ranges are infinite, then this does not raise a conceptual problem. The point, however, is that any biological population experiences only a vanishingly small range of the possible values of these variables. If the theory is ever to be connected to experimental data, this limitation is serious. It serves as a reminder that not only are **H** and **h** only properties of a population, they only are such properties for the particular range of environments that has been considered. In other words, high values for **H** or **h** in a particular experimental situation do not necessarily indicate that the mean value of a trait cannot be altered through the manipulation of environmental variables. Altering environmental variables beyond the current range might well increase (or decrease) the mean value of a trait even if **h** ≈ 1.[42]

APPENDIX 4.2

Analysis of the model of Table 4.4.1

The mean phenotypic value for the population, *m*, is given by:

$$\mu = p^2 d + 2pqh + q^2 r. \qquad (4.\text{A}2.1)$$

From Equation (4.A2.1) and the definitions of i, j, and k, using the facts that $p + q = 1$ and (therefore) $(p + q)^2 = p^2 + 2pq + q^2 = 1$,

$$p^2 i + 2pq j + q^2 k = 0. \tag{4.A2.2}$$

From the definition of a variance, the total phenotypic variance for the population, **V**, is given by

$$\mathbf{V} = p^2 i^2 + 2pq j^2 + q^2 k^2. \tag{4.A2.3}$$

Now if additive gene (allele) effects could account for the variation of phenotypic values completely, and if α is the contribution from A and β is the contribution from a, then the following equations would hold:

$$i = 2\alpha;$$
$$j = \alpha + \beta; \tag{4.A2.4}$$
$$k = 2\beta.$$

To account for as much of the phenotypic variance in terms of additive gene (allele) effects, α and β will be obtained by minimizing the sum of the mean squared deviations of the actual coded genotypic values from what would be obtained if additivity is perfect [that is, the Equations (4.A2.4) held]. Let $f(\alpha, \beta)$ be the sum of the mean squared deviations, that is,

$$f(\alpha, \beta) = p^2 (i - 2\alpha)^2 + 2pq(j - \alpha - \beta)^2 + q^2 (k - 2\beta)^2. \tag{4.A2.5}$$

To minimize $f(\alpha, \beta)$, set

$$\frac{\partial}{\partial \alpha} f(\alpha, \beta) = 0; \qquad \frac{\partial}{\partial \beta} f(\alpha, \beta) = 0. \tag{4.A2.6}$$

This gives:

$$p^2 (i - 2\alpha) + pq(j - \alpha - \beta) = 0; \tag{4.A2.7a}$$

$$pq(j - \alpha - \beta) + q^2 (k - 2\beta) = 0. \tag{4.A2.7b}$$

Adding Equations (4.A2.7a) and (4.A2.7b) and manipulating terms,

$$(2p^2 + 2pq)\alpha + (2q^2 + 2pq)\beta = p^2 i + 2pq j + q^2 k = 0. \tag{4.A2.8}$$

[The last equality follows from Equation (4.A2.2)]. From Equation (4.A2.8), using $p + q = 1$,

$$p\alpha + q\beta = 0. \tag{4.A2.9}$$

From Equations (4.A2.9), (4.A2.7a), and (4.A2.7b), once again using $p + q = 1$,

$$\alpha = pi + qj; \qquad \beta = pj + qk. \qquad (4.\text{A2}.10)$$

The *effect* of substituting the allele A for the allele a is defined as $\alpha - \beta$ and is, therefore, equal to $p(i - j) + q(j - k)$. Note that it depends on the frequencies of the alleles.

Finally, the additive genetic variance $\mathbf{V}_g$ is:

$$\mathbf{V}_g = 2(p\alpha^2 + q\beta^2). \qquad (4.\text{A2}.11)$$

(The factor of 2 arises from diploidy.) From Equations (4.A2.10) and (4.A2.11), after some algebraic manipulation,

$$\mathbf{V}_g = 2pq[p(i - j) + q(j - k)]^2. \qquad (4.\text{A2}.12)$$

In this model the total (phenotypic) variance, $\mathbf{V}$, is the sum of $\mathbf{V}_g$ and the variance due to the dominance of one trait over another, that is the dominance variance, $\mathbf{V}_d$. Therefore, using Equations (4.A2.3) and (4.A2.12):

$$
\begin{aligned}
\mathbf{V}_d &= p^2 i^2 + 2pqj^2 + q^2 k^2 - 2pq[p(i - j) + q(j - k)]^2 \\
&= p^2 q^2 [(i - j) - (j - k)]^2. \qquad (4.\text{A2}.13)
\end{aligned}
$$

The narrow heritability, $\mathbf{h}$ can now be specified by:

$$
\begin{aligned}
\mathbf{h} &= \frac{pq[p(i - j) + q(j - k)]^2}{pq[p(i - j) + q(j - k)]^2 + p^2 q^2 [(i - j) - (j - k)]^2} \\
&= \frac{[p(i - j) + q(j - k)]^2}{[p(i - j) + q(j - k)]^2 + pq[(i - j) - (j - k)]^2} \qquad (4.4.1)
\end{aligned}
$$

where $\mathbf{V}$ has been written in the form of $\mathbf{V}_g + \mathbf{V}_d$. (This will help the analysis that follows.) Note that because of Equation (4.A2.2) there are only two independent parameters among i, j, and k. Define $m = i - j$ and $n = j - k$. Then m and n may be taken to be independent and $\mathbf{h}$ can be written in the form

$$\mathbf{h} = \frac{[pm + (1 - p)n]^2}{[pm(1 - p)n]^2 + p(1 - p)(m - n)^2}. \qquad (4.4.2)$$

This form for $\mathbf{h}$ is particularly easy to analyze.

It should be obvious from either Equation (4.A2.14) or Equation (4.A2.15) that $\mathbf{h}$ depends on p, that is, the frequency of the A allele (or, alternatively, on $q = 1 - p$, the frequency of the a allele) in the population. Figure 4.4.4

shows this dependence. For all values of m and n, **h** shows some frequency dependence. For many, **h** varies a lot, going down from 1 to 0 and then returning to 1. The frequency of A for which **h** is equal to 0 can be easily calculated. Setting the numerator of Equation (4.A2.14) equal to 0, and using the definitions of m and n:

$$p = \frac{n}{n-m} = \frac{k-j}{i-2j+k}. \tag{4.A2.16}$$

Since $0 \le p \le 1$ (because p is a frequency), the only constraints are that

$$n > 0 \Rightarrow m < 0 \quad (\text{or } j > k \Rightarrow i < j); \tag{4.A2.17a}$$

$$n < 0 \Rightarrow m > 0 \quad (\text{or } j < k \Rightarrow i > j). \tag{4.A2.17b}$$

This can be interpreted as requiring that the genotypic value for the hetero-zygote must either be above or be below the genotypic values for both homozygotes. If $n \ne 0$, and $m \ne 0$, **h** can be equal to 1 in two ways: (i) if $m = n$, that is, that there is no dominance, which is not very interesting; and (ii) if $p = 1$ (or $p = 0$), that is, one (or the other) of the alleles is close to fixation. Moreover, as Figure 4.4.4 shows, for all values of m and n, **h** is close to 1 near these frequencies. Thus, a high value for **h** may only indicate that one of the alleles is rare in the population. If $n = 0$ (or if $m = 0$), **h** is equal to p (or $1 - p$). Therefore, when there is complete dominance, **h** gets progressively higher as the recessive allele increases in frequency in the population.

To see how **H** can be frequency-dependent, consider a second model that differs from the first only insofar as there is an environmental dependence of the phenotype, but $R = 0$ and $\mathbf{Cv}(G, E) = 0$ (using the notation of Appendix 4.1). Thus, the minimal interpretation is still tenable. For simplicity assume that $\mathbf{V}_E$ is a constant. Using Equation (4.1.3) and the value for $\mathbf{V}_d$ that was calculated above,

$$\mathbf{H} = \frac{[pm + (1-p)n]^2 + p(1-p)(m-n)^2}{[pm + (1-p)n]^2 + p(1-p)(m-n)^2 + \mathbf{V}_E}. \tag{4.4.3}$$

100

5

Reduction and classical genetics

If heritability analysis exhausted the techniques available to it for explanation, genetics would be in a sorry state. Luckily, the pursuit of genetic explanation has available an array of reliable techniques that underscore the irrelevance of heritability analysis to modern genetics. These techniques use known rules about genes, especially their transmission, to infer which traits can be explained from a genetic basis. The basic pattern for this type of explanation was set by Mendel (1866). Mendel hypothesized that many traits in his pea plants were controlled by pairs of inherited factors, now called "genes" or, more accurately "alleles." Each factor in a pair was one of two possible types. These factors obeyed definite rules during their transmission from one generation to the next: each member of a pair was transmitted independently of the other, and each pair was transmitted independently of other pairs. "Dominant" traits were those controlled by pairs of similar or different factors; "recessive" traits were controlled by pairs of similar factors. Mendel's rules for the transmission of factors established definite expectations for the inheritance of both dominant and recessive traits. In this sense, the factors explained those traits that satisfied these expectations. But they were silent about the mechanisms of phenogenesis, that is, how the factors brought about the traits.

Modern genetic explanation follows the same pattern. Accepting Mendel's laws (though with an important exception that will be discussed in detail in § 5.1), genes are said to be able to explain the origin of a trait if its pattern of inheritance is one that is predicted from those laws. Moreover, genes are also assumed to be capable of such an explanation if a trait's inheritance appears to be strongly correlated with one that is known to have a genetic explanation. These are the explanatory and inferential strategies that will be explored in this chapter. The **F**-rules are those that are supplied by Mendel's laws, as modified by the discovery of what is called "linkage" (see § 5.1 below) along with the critical assumption that genes can explain traits, but not the converse. What is to be explained is the origin of various phenotypic traits.

Since the transmission of genes turns out to be governed by statistical rules applicable to whole populations, this process of inference to a genetic explanation is necessarily statistical. Almost a century of research has made such inference remarkably sophisticated. These techniques long precede molecular biology though, today, they are often used in conjunction with molecular techniques. They have been developed from the statistical exploration of the Mendel's laws and their (partial) modification during the 1920s and 1930s. What makes these techniques particularly important in the present context is that they provide the basis for most of the claims that certain traits are genetic.

This chapter will discuss these inferential techniques in some detail. First, Mendel's laws, on which the traditional techniques have relied, will be stated (§ 5.1). Next, an example of reduction using Mendelian genetics that is conceptually important, but usually has been misrepresented by philosophers and historians of science, will be discussed (§ 5.2): this is the reduction of biometry to Mendelism, which led to the emergence of quantitative genetics in its modern form. It is philosophically interesting because it shows the extent to which epistemically questionable approximations may vitiate the value of a reduction. Following this, two of the traditional inferential techniques used in classical genetics for genetic explanation, segregation analysis (§ 5.3) and linkage analysis (§ 5.4), will be discussed. A series of technical caveats that have the potential to undermine these techniques will be analyzed in § 5.5. The nature of reduction in classical genetics will be discussed in § 5.6. Finally, the last section (§ 5.7) will briefly note some of the new directions genetic inference has taken since the 1980s. Their relevance to the problem of reduction will be tentatively assessed.

The strategy usually followed in all of these techniques, traditional or new, is model-fitting: assume a plausible model of gene transmission (including a specific number of loci and definite numbers of alleles at each of them).[1] Next assume that a trait is controlled by those numbers of loci and alleles. Then test whether the observed pattern of phenotypic transmission is consistent with that model. If not, try other models of gene transmission until the whole project becomes implausible. If some model is plausible, there is reason to believe that the type of genes assumed in the model can explain the origin of that trait: the trait, in this sense, is reduced to the genes. While this is the usual strategy, a caveat is necessary. In at least two cases genetic inference involves testing, at least partly, for a failure of a model rather than a fit to it: (i) linkage analysis (see § 5.4), in some sense, involves testing for a failure of a model (more specifically, the law of independent assortment), though since the actual extent of linkage is usually estimated, linkage analysis in practice is a model-fitting procedure; and (ii) the allele-sharing method (see § 5.7) entirely involves testing for failure (in this case of the law of segregation of alleles).

Genetic explanation, that is, when the origin of a trait is explained from a genetic basis, will be called "genetic reduction."[2] The thesis that these are ubiquitous, and the concomitant belief that they should be pursued, will both be called "genetic reductionism" (to distinguish it from another type of reduction and reductionism that will be discussed in the next chapter). This chapter, therefore, is about the strategies of genetic reductionism. The important characteristic of a genetic explanation or reduction is that genes alone provide the best explanation (in the given context) for the origin of the trait in question. This point is far too often ignored in the ongoing disputes between genetic reductionists and their critics. Merely pointing out that environmental factors have some role in the origin of a trait does not, by itself, destroy a potential genetic reduction. Indeed, as pointed out in Chapter 1 (§ 1.1), it is a trivial developmental fact that both genetic and nongenetic factors have some role to play in the origin of every trait. It still remains possible, however, that genes alone can explain all that there is to explain for the origin of some traits. In these contexts the genes alone bear the explanatory weight: citing genes suffices to explain the phenogenesis of traits even though nongenetic factors have a role in pheongenesis. The techniques discussed in this chapter will be judged according to their ability to do just that.

Finally, a disclaimer about the title of this chapter is necessary. Though the title refers to classical genetics (which is usually used to refer to all of premolecular genetics), this chapter confines itself to transmission genetics only. There are two reasons for this: (i) this book is oriented towards understanding and assessing the genetic explanation or reduction of traits. In that restricted arena, within classical genetics, it is transmission genetics that has played and continues to play the central explanatory role; and (ii) the rest of classical genetics (for instance, cytogenetics), which had a considerably lesser explanatory role, will be touched upon (though not treated in detail) in the next chapter, where it will be viewed as a stage in the creation of molecular genetics. Were this book an attempt to explicate all explanatory strategies used in every biological subdomain that has ever been at least partly described as "genetics," such a cursory treatment of cytogenetics and these other such subdomains of genetics would not be justifiable. In the restricted domain framed by issues connected with reductionism, it appears reasonable.

Genetics and reductionism

5.1. MENDEL'S LAWS

For expository ease, the discussions of this book are confined to diploid organisms, for which two laws that have come to be called Mendel's laws are relevant.[3] These were stated, though not emphasized, by Mendel (1866). This lack of emphasis has generated a controversy over whether Mendel discovered these laws.[4] This historical issue is beyond the scope of this book and will be ignored. What is more interesting in this context is that Mendel's laws can be used to infer the importance of genes for some phenotypic characteristics of organisms and thereby be used to explain them (see § 5.3 and § 5.4). Mendel's laws assume that for each gene an organism has two homologous loci (which can be responsible for the various phenotypic traits) and one of several possible alleles at each locus. That there are two homologous loci is a consequence of their being paired chromosomes. This is called diploidy. Haploid organisms have one such locus, polyploid organisms have more than two homologous loci. For diploids and generally for polyploids "locus" is routinely used in genetics to refer to all the homologous loci. This abbreviated use does not generate any genuine confusion once the context is specified. It will also be adopted here (and was used in Chapter 4) when no greater precision is necessary. In the statement of Mendel's laws that follow, for instance, "locus" will be used to refer to both homologous loci.

Both of Mendel's laws are assertions of statistical independence.

(i) *Law of Independent Segregation* (of Alleles): For each individual, each of the two alleles at a locus segregates to a germ cell with equal probability.

(ii) *Law of Independent Assortment* (of Alleles at Different Loci): The allele pairs at different loci assort independently of each other (when the germ cells are formed).

The status of each of these laws is radically different. Violations of the law of independent segregation, known as segregation distortion, are relatively rare. Consequently, model fitting using this law is a reasonable strategy, especially when only one locus is involved and there is no need to invoke the law of independent assortment. This is the fact on which segregation analysis (§ 5.3) is based. This strategy goes back to Mendel. In a sense, all that modern work has done is add statistical sophistication.

In sharp contrast, the law of independent assortment is routinely violated. The reason for this is that loci that physically are on the same chromosome tend to get inherited together.[5] Loci on different chromosomes assort independently. However, because homologous chromosomes exchange (corresponding) parts during meiosis (a phenomenon called "crossing-over"), there is a "recombination" of alleles. The probability of crossing-over is a monotonically increasing function of the distance between loci, at least for distances short enough that there is no more than one cross-over. Therefore, even alleles at loci on the same chromosome can assort independently provided that they are far apart. However, the failure of some loci to assort in this fashion can be used to detect a previously unknown genetic basis for some trait if it tends to get inherited together with a trait known to have a genetic basis. This is the strategy followed by linkage analysis (§ 5.4). The basic strategy is to use a failure of the law of independent assortment to infer a genetic explanation for some trait. However, as noted before, since the degree of linkage is usually simultaneously estimated, linkage analysis usually remains a model-fitting procedure even though it uses the failure of the law of independent assortment.

5.2. BIOMETRY AND MENDELISM

Before turning to an analysis of the strategies for genetic explanation that were outlined in the previous section, which are generally useful only for discrete or discontinuous traits, that is, those that only assume discrete values (such as eye color in animals, or seed color in Mendel's peas), this section will analyze the first attempt to reduce continuous traits (such as the stature of weight of organisms) to genetics. That attempt was mostly due to Fisher (1918) and, since it is regarded as having laid the foundations for quantitative genetics, is of considerable historical importance.

The mathematical exploration of heredity began with Mendel (1866), but his work remained unknown for a generation. Meanwhile, from entirely different assumptions, a mathematical theory that eventually came to be known as "biometry" was developed in the United Kingdom, thanks largely to Galton and Pearson.[6] Perhaps the most lasting contribution of the efforts of this school was the development of modern statistics, including the concept of correlation.[7] After Mendelism was rediscovered around 1900, in the United Kingdom a bitter dispute broke out between the Mendelians, led by Bateson, and the biometricians, led by Weldon and assisted by Pearson.[8]

The conventional view of the end of this dispute is that the scientific issues involved were only resolved through the work of Fisher, Haldane, and Wright who, by developing a mathematical theory of (Mendelian) population genetics, effected a "synthesis" between biometry and Mendelism.[9] There are many problems with this account of the history of evolutionary genetics. For instance, the dispute between the Mendelians and biometricians was confined to the United Kingdom, and Wright, working in the United States, showed little interest in it. Similarly, Haldane was largely uninfluenced by it. Of the three figures whose work allegedly led to a resolution of the dispute, only Fisher (1918) explicitly attempted to resolve it.

However, this attempt consisted of a putative reduction of biometry to Mendelism rather than a synthesis of the two. That will be the only point pursued here. Besides correcting the conceptual reconstruction of an important episode in the history of genetics, this example is particularly relevant to the themes explored in this book for two reasons: (i) the approximations introduced by Fisher were epistemically so dubious that the value of this reduction as an explanation is at best marginal; and (ii) this example provides a particularly clear picture of what genetic reductionism is supposed to be.

Classical biometry had a vigorous life of only about twenty years, from 1890 to 1910. It came under attack from the Mendelians even before it matured, and its principles were never systematically enunciated. The only relatively complete description of the biometrical principles is found in the second edition of Pearson's (1900) *The Grammar of Science*. The account given here draws heavily on that work. The striking difference between Mendelism and biometry was that, whereas the former studied discontinuous traits, biometry studied continuously varying traits. The biometricians did not generally doubt that Mendelism could explain the properties of discrete traits to some extent; however, as they correctly noted, "pure" Mendelism (that is, without linkage and with complete dominance, as in Mendel's original paper) was applicable only in rare cases. They doubted that the properties of continuous traits that for them were the vast majority of the interesting traits could be given Mendelian explanations.

For the continuous traits that they were interested in, the biometricians found three types of rules.

(i) By extensive empirical investigations they showed that a vast majority of continuous traits, including height and weight in humans and the number of leaves in trees, were normally distributed. Even when some traits were apparently not normally distributed, their observed distribution could be resolved into sums of two (or possibly a few more) normal distributions. This meant that the population being investigated could be subdivided into two (or more) distinct subpopulations in each of which the trait was normally distributed, though around a different mean (and possibly with a different variance) in each case.

(ii) The biometricians measured and compiled correlation coefficients for these traits for various relatives, including parent–offspring, sibling, and cousin correlations. This is where their quantification of the concept of correlation turned out to be most useful.

(iii) Finally, the central theoretical claim of the school was the "law of ancestral heredity." Stated in rudimentary form by Galton (1865), it was given
its most elaborate formulation by Pearson in the 1890s.[10] For some trait,
let h be its deviation in an individual from the mean for the population
of that generation, and let σ be the standard deviation. For ancestors of
these individuals i generations ago, let H_i be the deviation of the "mid-
parental" value (a type of weighted average between parents) from the
mean mid-parental value, and let Σ_i be the standard deviation (among
the ancestors).[11] Then, Pearson (for example, 1900) claimed

$$h = \sum_{i=1}^{\infty} \gamma_i \frac{\sigma}{\Sigma_i} H_i \tag{5.2.1}$$

where the γ_i are numerical quantities that depend on the correlation coefficients for that trait. From Galton's and his own analysis of available
data, Pearson concluded that the Σ_i could be replaced by the σ_i, the
standard deviations for the entire population in that generation, and that
the $\gamma_i = \gamma a^i$, with $0 \leq a \leq 1$, and with γ being some constant. Therefore, these coefficients decreased geometrically, and the contributions of
previous generations became progressively less significant. With these
substitutions Equation (5.2.1) becomes the law of ancestral heredity:

$$h = \gamma \sum_{i=1}^{\infty} a^i \frac{\sigma}{\sigma_i} H_i. \tag{5.2.2}$$

Pearson (1904a, b) attempted to derive these rules from Mendel's laws,
but only halfheartedly, and concluded that the two sets of rules were inconsistent. Moreover, the correlation coefficients measured in his laboratory
could not be obtained from those laws. Yule (1902, 1906) and Weinberg
(1908) provided more optimistic assessments, but a systematic exploration
was left to Fisher (1918) in a path-breaking paper, "The Correlation between
Relatives on the Supposition of Mendelian Inheritance." Fisher's crucial assumption was that the continuous traits were "determined by a large number
of Mendelian factors." This seems innocuous enough, but Fisher used it to
argue that the distribution must be normal. In effect, what he assumes is
that the number of factors is almost infinite, each has very little effect, and
that they act independently – the asymptotic normality of the distribution
is then a consequence of (one version of) the Central Limit Theorem for
distributions.[12]

Once the normality of the distributions was assured, Fisher calculated
the various correlations to be expected and found them in approximate

agreement with the measurements of the biometricians. Finally, he claimed to provide a derivation of the law of ancestral heredity from these assumptions [Fisher (1918), § 17]. However, what that derivation actually shows is only that the ancestral contributions decreased in a geometric fashion. There was no detailed agreement with Pearson's (1900) statement of the law, but this point was ignored in the subsequent literature. It was assumed after Fisher's (1918) work that there was no problem to be resolved between biometry and Mendelism. However, this work came to be regarded as a synthesis, as noted before, even though the very title of Fisher's paper clearly indicates that the relation between the two fields was not seen as symmetric. Rather, Mendelian rules were to bear the explanatory weight, as biometry was reduced to Mendelism.

As noted at the beginning of this section, there are two reasons why this example is important in the context of this book.

(i) The critical approximation that Fisher makes is his assumption that there are an infinite number of factors, each with negligible effect and acting independently. This is what permits his implicit use of the Central Limit Theorem. All three assumptions in this approximation are independently counterfactual: (a) that the number of factors is infinite; (b) that they are independent; and (c) that each has negligible effect. One consequence of these assumptions is that the details of the action of the Mendelian alleles is irrelevant. It is even irrelevant, in this argument, what the factors are – they could just as easily be environmental factors as genetic ones. However, this would require a more complicated calculation for the correlation coefficients.

That the traits have a normal distribution once these assumptions are made is, in a sense, purely an artifact of these assumptions. Consequently, Fisher's explanation on the "supposition of Mendelian inheritance" actually puts little explanatory weight on the Mendelian principles. At best what Fisher's argument does is to establish a weak form of consistency (or compatibility) between Mendelism and biometry; to that extent, Fisher addresses the reservations that Pearson (1904a, b) had about Mendelism. The point being made here, that Fisher's results establish little more than compatibility between biometry and Mendelism, is not entirely new. In 1957 Waddington, for instance, pointed out that one of the only two accomplishments of mathematical population genetics was that "it demonstrated that continuous variation, which is much commoner in nature [than discontinuous variation], does not differ essentially in its biological causation from discontinuous, but like it depends on genes[13] (1957, p. 61)." If the ontological tenor of these remarks is ignored, Waddington's observation is only slightly stronger than the claim that the laws governing the transmission of continuous traits are consistent with the principles of Mendelian genetics – he does leave open some possibility for explanation beyond the mere demonstration of consistency (or compatibility). This situation, that a putative reduction does not explain but only establishes weak consistency between two fields, is typical when systematically counterfactual approximations have to be made.[14] As a result of this unsatisfactory reduction, but helped no doubt by the consistency that could be claimed for biometry and Mendelism, quantitative genetics, which is what biometry (and biometrical genetics) slowly became, developed on its own paying little more than lip service to Mendelism.[15]

(ii) The biometrical rules, as Pearson (for example, 1900) repeatedly stated, were intended to describe the regularities that held during phenotypic transmission.[16] Pearson clearly expected that underlying genotypical rules would explain them. What the biometricians disagreed with was whether the Mendelian mechanisms were the proper explanation.[17] Thus, the entire scientific dispute was about explanation and, ipso facto, about reduction, rather than synthesis. In other words, what Fisher (1918) attempted to show was that the **F**-rules provided by Mendelism were sufficient for the explanation (and, therefore, reduction) of the properties of these continuous phenotypic traits just as they had been successful for discontinuous traits. The pattern of reasoning was the same as for the latter traits and, to the extent that continuous traits continue to be investigated genetically, this is the pattern that has persisted to this day. What is unfortunate, given Fisher's explicit aims, is that his explanation involved so many counterfactual assumptions. What has changed since 1918 is that Fisher's artificial approximations have been replaced by more detailed and specifically Mendelian assumptions. Fisher's approximations are, therefore, largely corrigible. A detailed analysis of the extent to which classical biometry can be recovered from Mendelism, or explained by it, without making epistemically problematic approximations remains to be undertaken. It would be philosophically interesting though probably of little scientific value.

5.3. SEGREGATION ANALYSIS

The reservations that have been expressed about Fisher's reduction of the laws of biometry to Mendelism indicate the relative insignificance of that

Table 5.3.1. *Mating table for segregation analysis with two alleles at one locus*

Parental genotypes	Offspring genotype frequencies			Offspring phenotype frequencies	
	A_1A_1	A_1A_2	A_2A_2	A	a
$A_1A_1 \times A_1A_1$	1	0	0	1	0
$A_1A_1 \times A_1A_2$	0.5	0.5	0	1	0
$A_1A_1 \times A_2A_2$	0	1	0	1	0
$A_1A_2 \times A_1A_2$	0.25	0.5	0.25	0.75	0.25
$A_1A_2 \times A_2A_2$	0	0.5	0.5	0.5	0.5
$A_2A_2 \times A_2A_2$	0	0	1	0	1

Note: The model assumes one locus with two alleles and random mating. The two alleles are A_1 and A_2. The genotypes are A_1A_1, A_1A_2, and A_2A_2. The phenotypes are A and a, with A being fully dominant over a (or, alternatively, A_1 is fully dominant over A_2). The first column lists the type of mating. The next three columns record the proportions of the various genotypes produced by each type of mating; these simply reflect the proportions of gametes produced of each (allelic) type. The last two columns take dominance into account and record the proportions of each phenotype among the offspring.

reduction qua reduction. The systematic use of Mendel's laws, however, allow two distinct patterns of reduction, one arising from each law. The law of segregation of alleles allows the use of pedigrees to infer a genetic explanation of traits. The law of independent assortment, though only through its violation, allows the use of linkage to make a similar inference: this will be discussed in the next section.

It is easy enough to demonstrate how segregation can be used to infer a genetic explanation of traits. For expository ease, the discussion here will be restricted to autosomal loci. In principle, only two steps are necessary.

(i) Mating tables corresponding to the various possible genetic configurations that might potentially explain the origin and transmission of a trait have to be constructed. Such a table gives the expected genotypic and phenotypic frequencies in the following generation from each type of mating (see Table 5.3.1 for an example);

(ii) The distribution of the traits in different pedigrees are compared with the frequencies expected on the basis of the table. This is a process of statistical inference and, therefore, comes with the usual philosophical and technical problems associated with statistical explanation. These are beyond the scope of this book and will not be discussed here.[18]

For a particularly simple example, which nevertheless illustrates the process of pedigree analysis accurately, consider a single autosomal locus, with two alleles, A_1 and A_2. The phenotypic trait A, which corresponds to the genotypes A_1A_1 and A_1A_2, is assumed to be dominant over the trait a, which corresponds to A_2A_2. The frequencies of the different genotypes and phenotypes that would arise, on the average, from various kinds of mating are shown in Table 5.3.1. (For other cases, even when attention is restricted to a single locus, different tables would have to be constructed. More than two alleles would introduce more genotypes: for instance, n alleles would require $n(n+1)/2$ genotypes potentially all distinguishable from each other. Even with only two alleles, if dominance is not complete, all three genotypes would be phenotypically distinguishable. Therefore, three phenotypes would have to be introduced and their expected frequencies would be the same as that of the genotypes in Table 5.3.1. Male X-linked traits would require a different table altogether.)

The intended use of this table is to compare phenotypic distributions observed in pedigrees across several generations with those that would be predicted from the last two columns. (For explicit examples, see (i)–(iii) below.) If the fit is good, in a statistical sense, then the inference is that the trait can be explained from the genetic basis, that is, from the rules for the transmission and expression of alleles that were used to construct Table 5.3.1. In this sense the phenotypic traits are reduced to the genes. If no table that can be constructed on the basis of such known genetic rules provides a good fit for the data, the attempted reduction fails. (The level of the genes is the **F**-realm in these reductions.)

Four historical examples will illustrate the use of segregation analysis. The first three are unproblematic, though the second lacks quantitative rigor. The fourth is intended as a warning; it illustrates a type of misuse of pedigree analysis that was typical during the heyday of eugenics.[19]

(i) Mendel's own experiments provide a striking illustration of how Table 5.3.1 could potentially be used. Since for each of the seven traits that he studied Mendel started with pure strains, the mating corresponds to the third line of the table. The F_1 generation (of hybrids) only consisted of A_1A_2 genotypes. The mating between these, to give rise to the F_2 generation, corresponds to the fourth line of the table. Since dominance was perfect for Mendel's traits, the A and a phenotypes are expected in a $3:1$ ratio. Historically, Mendel used this data to infer the existence both of such alleles (his "factors") and the law of segregation itself.

That law is usually not a focus of concern any more. Rather, if these patterns are recognized in a pedigree, then there is reason at least to suspect that genetic explanation of the trait is likely.

(ii) Mendel's use of segregation was a model of quantitative rigor and precision for his time. Many of the early uses of pedigree analysis lacked such precision. For instance, in 1902, shortly after the rediscovery of Mendel's laws, Garrod (1902) observed that "a very large proportion of [individuals with alkaptonuria] are children of first cousins... [Moreover, i]t will be noticed that among the families of parents who do not themselves exhibit the anomaly a proportion corresponding to sixty percent are the offspring of marriages of first cousins" (pp. 1617–18). From such meager data Garrod nevertheless concluded that "the law of heredity discovered by Mendel offers a reasonable account of such phenomena. . ... Such an explanation removes the question altogether out of the range of prejudice." He went on to conclude that alkaptonuria must be a recessive trait, a conclusion that was borne out by subsequent work.

(iii) The quantitative analysis of segregation in order to infer a genetic explanation seems to have begun, though at a very superficial level, with Bateson (1906) who explained the origin of brachydactyly and congenital cataract as dominant traits arising from two alleles at one locus. In the pedigree of a family founded by an individual with brachydactyly, all matings were with normal individuals, and the affected individual at the root of the pedigree had a normal parent. Therefore, the fifth line of Table 5.3.1 can be used, and there should be approximately equal numbers of affected and normal offspring. Bateson counted thirty-three normals and thirty-six abnormals, "a close approach to the expected equality" (p. 160). A similar argument established the genetic origin of congenital cataract.[20]

(iv) As late as 1927, following Davenport's (1915) earlier work, from data that allegedly showed that 30.4 percent of the family members of "nomadic delinquents" also showed "nomadism," whereas only 1.2 percent of the family members of "non-nomadic delinquents" showed that trait, Tinkle (1927) concluded that nomadism was hereditary. An investigation of three pedigrees then showed that the trait appeared to be "due to an independent, recessive factor." Apparently no further quantitative analysis was necessary to reach this remarkable conclusion. Tinkle did not stop there: nomadism was apparently not correlated with intelligence and was probably sex-linked (and only appeared in males).

Thus, the analysis of segregation in pedigrees, based on the rather remarkable power of the law of segregation of alleles, provided the first method of reducing traits to genes. Moreover, as is common for all good reductions (see Chapter 3, § 3.10), the success of these explanations provides a strategy for exploring other questions. In fact, this is the most prevalent contemporary use of segregation analysis, especially in human genetics: if the genetic basis for some trait is known, pedigree analysis can be used, for instance, to reconstruct geneologies (with paternity testing being the simplest example) or to identify the origin of specific alleles (Thompson 1986). The methods involved can be illustrated by a relatively simple example, due to Haldane (1938a), which analyzes the transmission of hemophilia – an X-linked recessive trait – in the royal families of Europe.[21]

Figure 5.3.1 [redrawn from Haldane (1938a)] is a pedigree showing descendants of Queen Victoria through male hemophiliacs or females. In Haldane's analysis the genetic pattern of inheritance of hemophilia was assumed to be known. All that Figure 5.3.2 does is verify the occurrence of hemophilia in the pedigree, judged phenotypically by the inability of blood to clot properly. Assuming that females are at most heterozygous for the hemophilia allele, their sons would exhibit the disease with probability 1, while their daughters would also be heterozygous with probability 0.5. (Therefore, the daughters' sons should exhibit the trait with probability 0.5.) The pedigree shows this pattern well within the usual statistical fluctuations. (Note, however, which females were carriers, that is, heterozygous for the allele, had to be inferred from their children, that is, from the pedigree itself.) The only uncertainty in the analysis arises because four male descendants of Victoria died in infancy, and there is insufficient evidence as to whether they were hemophiliacs. (Problems of this sort are standard in the analysis of pedigrees, and this is precisely the reason why increasingly sophisticated statistical techniques are necessary.)

The purpose of Haldane's analysis was not to show that hemophilia could be explained through genetics. That was well-known. Rather, the question that he asked was that of the origin of the disease (in the sense of the mutation responsible for the disease). This required an investigation of the ancestry of Victoria, which is shown in Figure 5.3.2. It turns out that neither her maternal uncles, nor her maternal grandfather, nor her half-brother were hemophiliacs. Nor are there any hemophiliacs in her mother's family or in the descendants of her maternal aunt. The source of the allele, therefore, must have been a mutation, and the mutation must have occurred in her father who was not himself a hemophiliac. As Haldane concludes: "The gene must have originated by mutation, and the most probable place and time where the mutation may have

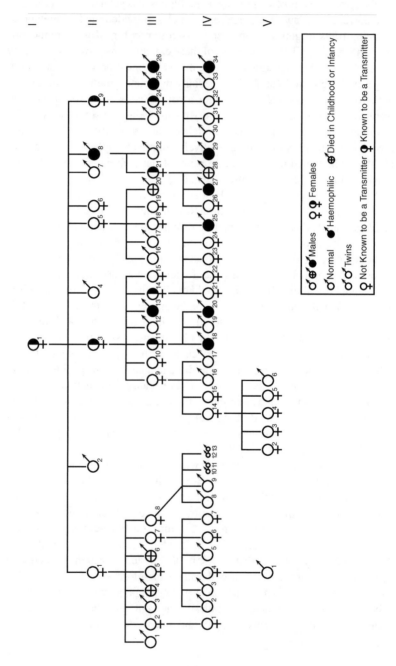

Figure 5.3.1 Descendants of Victoria through Male Hemophiliacs and Females [after Haldane (1938a)].

I (1) Victoria (b. 1819), Queen of England.

II (1) Victoria (b. 1840) × Frederick, later Emperor of Germany; (2) Edward VII (b. 1841), King of England; (3) Alice (b. 1843) × Prince Louis of Hesse; (4) Alfred (b. 1844), Duke of Edinburgh; (5) Helena (b. 1846) × Prince Christian of Schleswig-Holstein; (6) Louise (b. 1848) × Duke of Argyll; (7) Arthur (b. 1850), Duke of Connaught; (8) Leopold (b. 1853) × Princess Helena of Waldeck; (9) Beatrice (b. 1857) × Prince Henry of Battenberg.

III (1) William II (b. 1859), Emperor of Germany; (2) Charlotte (b. 1860) × Duke of Saxe-Meiningen; (3) Henry (b. 1862), Prince of Prussia; (4) Sigismund (b. 1864); (5) Frederica (b. 1866); (6) Waldemar (b. 1868); (7) Sophia (b. 1870) × Constantine, King of Greece; (8) Margaret (b. 1872) × Frederick, Duke of Hesse-Cassel; (9) Victoria (b. 1863) × Prince Louis of Battenburg (Marquess of Milford Haven); (10) Elizabeth (b. 1864) × Grand Duke Sergius of Russia; (11) Irene (b. 1866) × Prince Henry of Prussia [III (3)]; (12) Ernest (b. 1868), Grand Duke of Hesse; (13) Frederick William (b. 1870); (14) Alexandra (b. 1872) × Nicholas II, Tsar of Russia; (15) Mary Victoria (b. 1874); (16) Christian Victor (b. 1867); (17) Albert (b. 1869); (18) Victoria (b. 1870); (19) Louise (b. 1872); (20) Harold (b. 1876); (21) Alice (b. 1883) × Earl of Athlone; (22) Charles Edward (b. 1884), Duke of Albany; (23) Alexander (b. 1886), Marquess of Carisbrooke; (24) Victoria Eugenie (b. 1887) × Alfonso XIII of Spain; (25) Leopold (b. 1884), Lord Mountbatten; (26) Maurice (b. 1891), Prince of Battenberg.

IV (1) Feodora Maria (b. 1887) × Henry XXX of Reuss; (2) George (b. 1890), King of Hellenes; (3) Alexander (b. 1893), King of Hellenes; (4) Helena (b. 1896) × Carol of Roumania; (5) Paulos (b. 1901); (6) Irene (b. 1904); (7) Catharine (b. 1913); (8) Frederick Wilhelm (b. 1893), Prince of Hesse; (9) Maximillian (b. 1894); (10) Philipp (b. 1896); (11) Wolfgang (b. 1896); (12) Richard (b. 1901); (13) Christoph (b. 1901); (14) Victoria (b. 1885) × Prince Andrew of Greece; (15) Louise (b. 1889) × Crown Prince of Sweden; (16) George (b. 1892), present Marquess of Milford Haven; (17) Louis (b. 1900); (18) Waldemar (b. 1889), Prince of Prussia; (19) Sigismund (b. 1896); (20) Heinrich (b. 1900); (21) Olga (b. 1895), Grand Duchess; (22) Tatiana (b. 1897), Grand Duchess; (23) Marie (b. 1899), Grand Duchess; (24) Anastasia (b. 1901), Grand Duchess; (25) Alexis (b. 1904), Tsarevitch; (26) May (b. 1906) × Captain Henry Abel-Smith; (27) Rupert (b. 1907); (28) Maurice (b. 1910); (29) Alfonso (b. 1907), Prince of Asturias, now Count of Covadonga; (30) Jaime (b. 1908); (31) Beatrice (b. 1909); (32) Maria (b. 1911); (33) Juan (b. 1913); (34) Gonzalo (b. 1914).

V (1) Michael (b. 1921), Prince of Alba-Iulia; (2) Margaret (b. 1905) × Godfrey of Hohenlohe-Langenburge; (3) Theodora (b. 1906) × Margrave of Baden; (4) Cecilia (b. 1911) × Grand Duke of Hesse; (5) Sophia (b. 1914) × Christopher of Hesse.; (6) Phillipos (b. 1921), Prince of Greece.

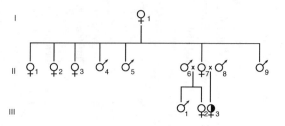

Figure 5.3.2. Ancestry of Victoria [after Haldane (1938a)]:

I (1) Augusta, Princess of Reuss, (b. 1757) × Frederick, Duke of Saxe-Coburg-Saalfeld.

II (1) Sophia (b. 1778); (2) Antoinette (b. 1779); (3) Henrietta (b. 1781); (4) Ernest, Duke of Saxe-Coburg-Gotha, (b. 1784); (5) Ferdinand, (b. 1785); (6) Emich, Prince of Leiningen; (7) Victoria (b. 1786); (8) Edward, Duke of Kent (b. 1799); (9) Leopold, King of the Belgians (b. 1790).

III (1) Charles, Prince of Leinengen (b. 1804); (2) Anne (b. 1807); (3) Victoria (b. 1819), Queen of England.

occurred was in the nucleus of one of the testicles of Edward, Duke of Kent, in the year 1818" (p. 135). The precise date, however, is conjectural; the evidence only indicates that the mutation occurred by 1818.

5.4.   LINKAGE ANALYSIS

Almost as soon as Mendelism was rediscovered around 1900, it was realized that the law of independent assortment was not universally valid.[22] This phenomenon, which was initially called "reduplication," eventually came to be called "linkage" after Morgan (1910) interpreted it as a result of loci being on the same chromosome.[23] Loci that tend to be inherited together (in violation of the law of independent assortment) belong to the same "linkage group." The physical interpretation of membership of the same linkage group is that these loci are on the same chromosome. Consequently, they tend to be inherited together. However, because of "crossing-over," that is, the exchange of parts between homologous chromosomes during cell division, alleles at loci on the same chromosome may yet be separated and, thereby, satisfy the law of independent assortment. This is called the "recombination" of alleles.

However, while this physical interpretation was critical to the historical development of linkage analysis, strictly speaking, it is irrelevant to an inference to a genetic explanation via linkage analysis. What matters,

116

simply, is the possibility of representing loci as organized in disjoint linkage groups, no matter how these groups are physically constituted. Similarly, it turns out that the loci within each group can usually be arranged in an ordered linear set with a definite distance separating each pair of loci. The physical interpretation of this phenomenon is that each locus corresponds to a particular position on a chromosome. However, once again, the physical interpretation, however scientifically important and interesting, is not necessarily for the purpose of genetic explanation. All that matters is that the loci that form a linkage group and tend to be inherited together are organized in such a set. The significance of this point, that the physical interpretation is only incidental to a genetic explanation, will be discussed in § 5.6.

In general, linkage is important for two reasons: (i) it permits an inference to a genetic explanation in many circumstances; and (ii) it enables "linkage mapping," the process of placing linked loci in the different linkage groups in a linear order and with a distance assigned between each pair of loci. The latter use of linkage analysis has historically been the more important. It was initiated by Sturtevant (1913) along with linkage analysis itself. Sturtevant studied six sex-linked traits in Drosophila that had each been shown (by Morgan and various collaborators) to be under the control of one locus. Accepting Morgan's physical interpretation of linked loci as positions on a chromosome, Sturtevant observed that recombination of alleles would result from cross-overs between chromosomes. Hypothesizing that the probability of a cross-over would be a monotonic function of the physical distance between loci on a chromosome, Sturtevant argued that

> the proportion of a 'cross-over' could be used as an index of the distance between any two factors. Then, by determining the distances... between A and B and between B and C, one should be able to predict AC. For, if proportion of cross-overs really represents distance, AC must be approximately, either AB plus BC, or AB minus BC, and not any intermediate value.... By using several pairs of factors one should be able to apply this test in several cases. (pp. 43–44)

Sturtevant's six loci passed the test and, on this basis, he constructed a map of the Drosophila sex chromosome in which (relative) positions were assigned to these six loci. In effect, Sturtevant introduced the first "mapping function," that is, a function that generates distances between loci from observed frequencies of recombination. Sturtevant's mapping function was trivially simple: the distance between loci was identified with the proportion of cross-overs (the "recombination fraction"). This function is obviously limited in its applicability: Since it ignores the possibility of multiple cross-overs between chromosomes, it can at best be used only for short distances. Starting with the work of Trow (1913) and continuing to the present, more sophisticated mapping functions have continued to be introduced.[24]

The aim in these endeavors is to place loci at precise positions in definite linkage groups. This, in turn, allows two tasks to be accomplished.

(i) Suppose that some trait is known to be under the control of a single locus. Mapping shows on which chromosome that locus is and, to varying degrees of precision (depending on the resolving power of the techniques used and the quality of the data), shows the exact position of the locus on that chromosome.[25] This is often a critical intermediate stage in the identification of the alleles involved in the genesis of a trait. The next step is a biochemical characterization of the alleles. (In most known cases, the sequence of DNA bases at that region of the chromosome is sufficient.) This allows further work: the determination of how that allele acts at the biochemical level and how the trait is brought about.

(ii) Systematic linkage mapping allows maps of entire chromosomes and, ultimately, entire genomes to be constructed. This is the genetic analog of anatomical study, a potentially useful description of an organism's genetic constitution.

In the present context, the first reason for pursuing linkage analysis – the inference to a genetic explanation – is more important than linkage mapping. What is required is the detection of linkage. If all that is of interest is whether the origin of a trait can be explained from a genetic basis, it is sufficient to show that there is an allele (or several alleles at different loci) responsible for it. It does not matter on which chromosome the locus is or, for that matter, how the alleles act at the biochemical level. These additional features are undoubtedly scientifically interesting but are not strictly required for a genetic reduction. Therefore, from this restricted point of view, there is nothing that linkage analysis accomplishes that goes beyond segregation analysis. However, the known statistical methods of linkage analysis are more powerful than those of segregation analysis.[26] Because of this, and because it furnishes the additional information mentioned before, as a strategy, linkage analysis is often preferred over segregation analysis.

Both the detection of linkage and linkage mapping have been pursued with increasing statistical sophistication. The statistical issues that have been raised include disputes about the relative values of Bayesian and non-Bayesian inference, as well as about the relative merits of parametric and nonparametric methods.[27] Though these are of significant philosophical interest and have not so far received the attention that they deserve, they are beyond the scope of this book and will be ignored. Similarly, further discussion of linkage mapping will be confined to the examples.[28]

Linkage detection consists of two steps.

(i) Linkage analysis (for whatever purpose) begins with the selection of some trait that is known to be inherited in accordance with the law of segregation of alleles. (Thus, segregation analysis must, in this sense, precede every use of linkage analysis.) Usually, this trait is under the control of a single locus. It is then hypothesized that the trait being investigated is linked to it.[29]

(ii) An attempt is then made (using a set of pedigrees) to detect the linkage between this trait and the one being investigated by looking for violations of the law of independent assortment. Thus, detecting a violation of a model of inheritance that assumes this law suffices for linkage detection. If there is such a violation, then the second trait is potentially inherited with the first one, and it is inferred that alleles at some locus linked to the first can explain the occurrence of the second trait. However, in practice, an attempt is also made to estimate the (relative) position of the second locus (as can be obtained from the recombination fraction). It is assumed that the greater the degree of linkage (in a statistical sense), the more reliable this explanation is.

The basic strategy of linkage detection will be illustrated using a simple model. This model considers only two loci since linkage analysis generally proceeds by considering two loci at a time.[30] For simplicity, it will also be assumed that there are only two alleles at each locus; extension to multiple alleles is straightforward. Let A_1, A_2 and B_1, B_2 be the two alleles at the **A**- and **B**-locus, respectively. If the individual's genotype is homozygous at even one locus, recombination will make no difference as far as the frequencies of the gametes (and, therefore, of the genotypes in the next generation) are concerned. If, however, the genotype is heterozygous at both loci, recombination will make a difference. Let the recombination fraction, that is, the probability that a gamete is a recombinant, be r. (Physically, this corresponds to the proportion of cross-overs.) If there is no linkage, that is, there is independent assortment, $r = 0.5$; if there is complete linkage, that is, there is no recombination, $r = 0$. Two cases – called "*cis*-" and "*trans*-," respectively – of the double heterozygote genotype must be distinguished: $A_1 B_1/A_2 B_2$ and $A_1 B_2/A_2 B_1$ where the two alleles on the same side of "/" are assumed to be in the same linkage group.[31] Table 5.4.1 shows

Table 5.4.1. *Gamete frequencies for linkage analysis*

	Gametes segregating to offspring			
Parent genotype	A_1B_1	A_1B_2	A_2B_1	A_2B_2
A_1B_1/A_2B_2	$0.5(1-r)$	$0.5r$	$0.5r$	$0.5(1-r)$
A_1B_2/A_2B_1	$0.5r$	$0.5(1-r)$	$0.5(1-r)$	$0.5r$

Note: The model assumes two loci with two alleles at each of them. Only the *cis-* (A_1B_1/A_2B_2) and *trans-* (A_1B_2/A_2B_1) double heterozygote parent genotypes are relevant. The probability that two alleles in one linkage group (chromosome) separate is the recombination fraction r. If they assort completely independently, $r = 0.5$. If they are linked, $0 \leq r \leq 0.5$, with $r = 0$ when linkage is complete. The four last columns give the frequencies of each type of gamete produced as a result from the parent genotype in that row. To detect linkage, one looks at the frequencies of offspring genotypes. The expected frequencies of offspring from *cis-* and *trans-* matings are obtained by multiplying the gametic frequencies.

the expected relative frequencies of the gametes expected from these two genotypes. The genotypic frequencies expected in the next generation are obtained by multiplying the gametic frequencies.

The expected frequencies for various genotypes among the offspring from matings involving such a parental genotype can now be calculated by simply multiplying these gametic frequencies with those from other parental genotypes (see the discussion of Table 5.3.1). By investigating a pedigree with at least one double heterozygote, r can be estimated using standard statistical techniques. If $r \neq 0.5$, there is linkage that gets tighter as r gets lower.

Four examples will show the use of linkage analysis. Only the last two of these provides an example of genetic explanation via linkage analysis. In the other cases a genetic explanation had already been inferred by segregation analysis. This should underscore the point that the usual use of linkage analysis is for mapping traits known to be controlled by some locus, rather than for linkage detection. The third also shows how linkage analysis may be an unreliable route to genetic explanation, especially in the case of human traits (which seem to excite most of the public attention).

(i) Following a strategy similar to that of Sturtevant (1913), by analyzing six traits that were sex-linked and, therefore, likely to be in the linkage group associated with the X-chromosome, Haldane (1936a) constructed the first map of a human chromosome – see Figure 5.4.1.

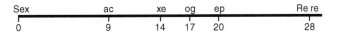

Figure 5.4.1. Provisional Map of the Human X Chromosome [after Haldane (1936a); by permission of "Nature"]. This was the first map of a human chromosome to be constructed. The mapping function used simply identified map distance with the proportion of recombinants. ac: Achromatopsia (total color blindness), recessive; xe: Xeroderma pigmentosum (severe light sensititvity of the skin), recessive; og: Oguchi's disease (night blindness with golden retinal pigmentation), recesive; ep: Epidermolysis bullosa dystrophica (skin disease), recessive; re: Retinitis pigmentosa without deafness, recessive; RE: Retinitis pigmentosa without deafness, dominant; some pedigrees only.

The "sex-damning" locus is taken to be the origin in this map, and the relative distances are established purely on the basis of r using Sturtevant's mapping function rather than more sophisticated functions that Haldane and others later devised. In effect, this means that the distance is identified with r as estimated from a table similar to Table 5.4.1. (That table cannot be directly used in this case because the traits in question are sex-linked, whereas that table was constructed for traits controlled by autosomal loci.)

(ii) The same kind of direct calculation led Bell and Haldane (1937) to infer linkage between hemophilia and color-blindness. They investigated six pedigrees in which hemophilia was present and calculated the expected probability of the occurrence of both color blindness and hemophilia in individuals in these pedigrees if the two were not linked. They found that the probability that the loci were unlinked was less than 4×10^{-6}. Segregation analysis (as discussed in the previous section) was sufficient to infer a genetic explanation for both hemophilia and color-blindness. Therefore, historically this determination of linkage also does not lead to a new genetic explanation. However, if only one of them had been known to have such an explanation, such an explanation could be inferred for the other.

(iii) In some families of the Amish community in Pennsylvania (USA), manic depression (bipolar affective disorder) appeared to be inherited. In 1987 a linkage was reported between this trait and the H-*ras* locus on the short arm of chromosome 11.[32] This appeared to be a case of genuine genetic explanation: the detection of linkage provided the reason for believing that manic depression was to be explained from the genes. However, studies of other pedigrees failed to confirm this result.[33] Finally, additional information on the original pedigree, when two previously unaffected individuals succumbed to manic depression, undermined the statistical basis for the assertion of linkage in the first place.[34] However, positive reports were routine, apparently linking manic depression to fourteen different loci. But, as Risch and Botstein (1996) note: "None of the 14... regions has been convincingly, or at least consistently replicated, although some regions have been implicated more than once." The controversy is unlikely to end soon. This story illustrates how linkage analysis can be used for genetic explanation but, more importantly, it also illustrates an important pitfall in the pursuit of genetic explanation via this route for organisms such as humans that are not subject to routine experimental manipulation to generate impeccable statistics. The statistical nature of these inferences requires that unless many different pedigrees have been used, any claim of genetic explanation be offered with considerable caution.

(iv) In all of the examples above, linkage analysis begins with the selection of a phenotypic trait known to be under the control of one locus. One can avoid this step by simply starting from a locus somehow specified on a chromosome, for instance, by a specific DNA sequence. (One can then regard this sequence as the phenotypic trait for the purpose of following the algorithm for linkage analysis that was outlined above.) The remarkable degree of DNA polymorphism in eukaryotes allows unique sequences to be often associated with a definite locus. This is the strategy behind restriction fragment length polymorphism (RFLP) linkage mapping.[35] Individual (potentially highly variable) DNA sequences are selected using random DNA probes that hybridize to them. Thus, each such probe defines a locus. The detection of tight linkage ("tight" in a statistical sense) to such a locus not only provides compelling evidence for a genetic explanation, it allows mapping onto physical regions of a chromosome with a very fine resolution. The first human locus to be found using this technique was that for Duchenne muscular dystrophy (on the X chromosome).[36] Another important early use was the identification of the locus for Huntington's disease (on the terminal band of the short arm of chromosome 4). In both these cases there was no scope for genetic explanation. From segregation analysis it was known that these traits were under the control of single loci. However, in the case of adult polycystic kidney disease the genetic explanation was not quite so clear prior to the identification of a locus on the short arm of chromosome 16. In this case linkage mapping came hand-in-hand with linkage detection and helped provide a reliable genetic explanation for the occurrence of the disease trait. This is the customary situation when linkage analysis is used for inferential/explanatory purposes.

From the point of view of genetic explanation, linkage analysis has two disadvantages when compared to segregation analysis: (i) one has to start with a trait known to be under the control of a single locus, and at this stage segregation analysis must be used; (ii) detecting linkage provides only indirect evidence that there are alleles at some locus that can explain the trait. It does not identify the locus, and the statistical inference that such a locus exists is often weaker than in segregation analysis. This is true even though, as noted before, the known statistical techniques for linkage analysis are more sophisticated than those for segregation analysis. The case of Huntington's disease illustrates the difficulty of finding the locus, and the alleles at that locus, even after the establishment of tight linkage. Though tight linkage to a locus on the short arm of chromosome 4 was shown as early as 1983, even using the most sophisticated techniques devised so far, it took ten years to identify the locus for the disease allele.[37]

However, linkage analysis has two advantages over segregation analysis: (i) when linkage is correctly detected, not only does it point to a genetic explanation, it also identifies the linkage group and, therefore, the chromosome on which that locus is; (ii) because of various factors – including the sophistication of the available statistical techniques and the various caveats that are discussed in the next section (§ 5.5) – linkage analysis can be pursued in many situations where segregation analysis is impotent.

5.5. CAVEATS

Both for explaining a trait from a genetic basis and for using segregation and linkage analyses to answer other questions, there are at least eight potentially confounding factors. Ignoring these, that is, assuming that they have negligible impact, involves approximations. In the case of segregation analysis and, to a lesser extent, in the case of linkage analysis, the problems with the first two of these (sampling and multiple loci) can be, and, in practice, have been, ameliorated through technical developments. There is no similarly complete technical solution for any of the other six and one of them, variable expressivity, cannot be incorporated into the formalism of segregation or linkage analysis. Therefore, while all these approximations, except variable expressivity, are corrigible in principle, only the first two are significantly corrigible in practice. If reduction is the aim, incomplete penetrance, genetic heterogeneity, and allelic homogeneity do not hinder genetic explanation, provided that they are interpreted properly. Variable expressivity, the existence of phenocopies, and nongenetic (e.g., cultural) modes of transmission can result in spurious claims of genetic explanation. These factors will now be discussed in some detail.

(i) Mendel's laws are inherently statistical, and any conclusion drawn from them must also be statistical. Restricting attention to the problem of genetic explanation, for segregation analysis, what has to be computed is the probability that the observed phenotypic distribution would occur, given the geneologies (or pedigrees, which are known), the hypothesized genetic model for inheritance, and the assumed relationship between genotype and phenotype.[38] This is obviously a problem of statistical inference with the usual technical problems, such as sampling bias, and interpretive ones, such as that of connecting the calculated probabilities to epistemological parameters (such as likelihood or degrees of belief).[39] However, segregation analysis in humans (and other populations not subject to experimental manipulation) involves an additional problem because except in rare cases the populations (that is, pedigrees) studied are not a random sample of the general population. Rather, a pedigree comes into attention because of the presence of a trait in it, or perhaps because of the extent to which it is present, or perhaps even because of how it is distributed. This situation requires a careful treatment of the sampling problem and, starting with the classical work of Hogben (1931), Fisher (1934a), and Haldane (1938b), technical methods have been systematically developed to resolve it.[40] For linkage analysis, the problems are similar when pedigrees have to be selected for use. The general statistical problems are also the same.

(ii) Were the law of independent assortment generally true, then the extension of segregation analysis to the study of traits dependent on multiple loci, or the simultaneous study of multiple traits, would be straightforward. At the other extreme, if there were no recombination at all, the extension of segregation analysis to multiple loci would be even more trivial.[41] Ordinary linkage with $0 < r < 0.5$, where r is the recombination fraction, introduces complications that, however, are only technical. Some progress in using segregation analysis for multiple loci in these circumstances has been made (Thompson 1986). In the case of linkage analysis, if a trait is under the control of several loci, all of these would have to be linked to the original locus. This raises severe technical difficulties. One approach is simply to study each of the linked loci independently; the problems with this are those connected with ordinary (two-locus) linkage analysis. However, the phenotype might have low penetrance and expressivity because there are other loci that may be involved. Problems of this sort will be discussed under (iii)–(vi) below. Another approach is to attempt what is called "multipoint" linkage analysis. There have been few methods developed toward this end, and the technical difficulties seem extraordinary.[42]

(iii) In general, tables such as Table 5.3.1 or 5.4.1 can be used only if the relation between genotype and phenotype is relatively straightforward. In general, the situations that are simplest to analyze are those in which either all genotypes are phenotypically distinguishable (e.g., in Table 5.4.1), or there is complete dominance (e.g., in Table 5.3.1) and, moreover, every individual with a particular genotype exhibits the corresponding phenotype while that phenotype is not exhibited without that genotype (that is, there are no phenocopies). All the examples given above satisfy these requirements. The first type of complication in the genotype-phenotype relation is due to *incomplete penetrance*. The penetrance of a phenotype x, given a genotype g, is defined as the conditional probability $p(x \mid g)$ of x being exhibited by an individual with genotype g in that population. If this probability is equal to 1, the penetrance is complete; otherwise it is incomplete.[43] The examples treated above assumed complete penetrance. However, incomplete penetrance is certainly not rare. Even for an alleged "genetic disease," diabetes mellitus, Rotter (1981) estimated penetrance to be only between 2 and 20 percent. When penetrance is incomplete, an individual with a certain genotype may not be identified as having that genotype. For segregation analysis this means that the computed probability of the observed phenotypic distribution will be lower than what would have been the case had the penetrance been complete. For linkage analysis this means that linkage may not be detected or may be detected with a lower statistical significance than would have been the case were penetrance complete.

Some care must be taken in assessing the significance of this situation. If segregation or linkage analysis is being used not for the sake of genetic explanation, but to answer some other question (such as the reconstruction of a genealogy or for the mapping of the loci), incomplete penetrance is a hindrance and has to be incorporated into the statistical techniques being used. If, however, genetic explanation is the aim, then the lower probability for a pedigree or for linkage, which would hurt the prospect of judging the explanation as successful, is entirely appropriate: incomplete penetrance indicates that there are other factors besides the given genotype that are relevant to the genesis of a trait. Therefore, the genotype cannot be the full explanation of why the trait occurs.

(iv) A trait may have *variable expressivity*, where expressivity is defined as the degree to which a trait is exhibited, given the genotype. Since both segregation and linkage analyses treat all individuals with a trait equally, variable expressivity is simply not incorporated into the statistical analyses that are used. It does not, therefore, affect the computed probability of an observed phenotypic distribution. This means that the use of these analyses for nonexplanatory purposes is unaffected. However, for genetic explanation, variable expressivity remains troublesome. It indicates that the genotype tells only part of the story that should be told for the origin of a trait. Moreover, if the variability of the expressivity includes low values, it suggests that the trait may be incompletely penetrant, and this raises even more problems for assessing the success of the explanation.

(v) If *phenocopies* exist, that is, if there are individuals who exhibit a phenotype associated with a given genotype for some reason other than the possession of that genotype, then the relation between genotype and phenotype may become many-one. Phenocopies result in false positives and are particularly prevalent when the phenotype is complex or ill-defined, as in the case of many diseases and most behavioral traits.[44] It cannot be stated a priori whether and to what extent the prevalence of phenocopies would lead to spurious genetic explanation through segregation or linkage analysis. However, it is quite easy to envision circumstances where this may occur, for instance, if penetrance is low and the trait is relatively rare. There is no technical solution to this problem, that is, one that provides a systematic scheme for taking all cases of phenocopy occurrence into account.

(vi) Even if a trait should be entirely explained from a genotypic basis, it may be the case that the same phenotype can arise from a variety of alleles at many different loci. This is known as *genetic* or *locus heterogeneity*. For instance, in humans the disease retinitis pigmentosa, which involves retinal degeneration (the phenotype), can arise because of mutant alleles at 14 different loci.[45] This will obviously hinder linkage detection unless it is recognized. However, it is unlikely to lead to spurious genetic explanation.

(vii) As noted in the previous section (§ 5.4), to distinguish double heterozygotes, linkage analysis requires at least that the two alleles be distinct at each locus. If this is not the case because of a high frequency of a particular allele (*allelic homogeneity*), traditional linkage

analysis will fail to detect linkage. As in the case of incomplete penetrance, there are ways to correct for this problem when it is recognized.
Moreover, it will potentially only lead to a situation where a genetic
explanation is incorrectly not inferred; it will not lead to spurious
explanation.

(viii) Finally, neither segregation nor linkage analysis considers the possibility that there may be patterns of nongenetic inheritance, including cultural inheritance, that would mimic genetic inheritance. For
complex traits, such as behavioral traits in humans, this could easily
lead to spurious genetic explanations. For instance, if mothers tried
to bring up sons to resemble maternal grandfathers, the phenotype
could at least show incomplete X-linkage. In general, segregation
and linkage analyses simply assume that scenarios like this are unlikely.

These complications show that there is a significant possibility that segregation or linkage analysis will lead to spurious genetic explanations. Moreover, these complications do not exhaust all the problems that may arise
when segregation or linkage analysis is used for genetic explanation. For
instance, in the case of allegedly hereditary nomadism, the problem, rather,
is in the definition of a trait. Similarly, it should not be forgotten that when
a single or at most a very few pedigrees are used, all that the analysis can
do is provide a genetic explanation of a trait in those pedigrees. To explain
that trait in general requires more work. These two points will be taken up
at the end of § 5.6.

5.6.   REDUCTION

That segregation and linkage analyses, when they are used for the purpose of
explaining features of phenotypic traits, do so from a genotypic basis should
be clear from what has been said. Thus, the fundamentalist assumption of
Chapter 3 (§ 3.2) is satisfied but, once again, as in Chapter 4, there are
some features of this assumption that are only implicit. What is asserted is
the epistemological primacy of the genotypic realm (the **F**-realm for these
reductions) over the phenotypic realm. The basic assumption is that the
genotype can mold the phenotype, but not vice versa. This assumption can
be further analyzed into two components.

(i) The rules of phenotypic transmission (from one generation to the next)
are governed – or regulated – by the rules of genotypic transmission, but
not vice versa. More precisely, the explanation of the rules of phenotypic

transmission are to be sought in the rules of genotypic transmission, but not vice versa. The rules of genotypic transmission are the law of segregation, necessary for both segregation and linkage analyses, and the law of independent assortment, as modified by linkage, necessary for linkage analysis.

(ii) Phenotypic features, especially phenotypic differences, are to be explained (as far as possible) from genotypic features, including differences, but not vice versa. One implication of this assumption is that phenotypic differences, if these do not arise from prior genotypic differences, cannot lead to genotypic differences. Of course, as has been noted in the previous section (§ 5.5), due to incomplete penetrance, variable expressivity, and other factors, the relation between genotype and phenotype is far from simple.

In explanations using either segregation or linkage analysis, the abstract hierarchy assumption of Chapter 3, § 3.2 is also satisfied, though the hierarchical structure assumed for the genome is trivial. Each locus permits two alleles (from a variety of possible types) in it.[46] A set of loci (with the alleles at them) forms a linkage group. A set of linkage groups forms the genome. What is less clear (but important) is that, in these explanations, it is irrelevant whether the hierarchically structured genome is a hierarchy of objects in physical space: The explanations and arguments that are used in pedigree or linkage analysis do not make the spatial hierarchy assumption of Chapter 3, § 3.2. Thus, the reductions that are the concern of this chapter are of the abstract hierarchical type [type (c) from Chapter 3, § 3.2]: they are not strong reductions.[47]

It is not being suggested here that the genome is not hierarchically structured in physical space, or that that hierarchy (of physical positions on chromosomes) is any different from what may be called the "genetic hierarchy" of linkage groups, loci, and alleles. To a rather remarkable degree of resolution, the two hierarchies are, indeed, the same.[48] The point, however, is that this fact is not used in either segregation or linkage analysis. In the case of segregation analysis using single, completely linked, or unlinked loci, this should be obvious: the genome is represented as sets of paired alleles that segregate during reproduction. In the case of linkage analysis, however, it appears that the linear arrangement of loci on chromosomes is critical to the analysis. Certainly, as noted in § 5.4, this explicit physical model guided the historical development of linkage analysis. However, all that is necessary for linkage analysis to work is that there exist linkage groups, with internal relations between members of the group that allow them to be placed in a linear (metrical) order using any of the mapping functions. Moreover, for linkage detection, which is all that is strictly required for genetic explanation, the internal structure of a linkage group is irrelevant. All that is required is that a trait is inherited according to the known rules for gene transmission. In linkage analysis one has available a known locus. All that has to be shown is that the trait being investigated can be accounted for from the hypothesis that there is a locus for it and that this locus belongs to the same linkage group as the known one.

Perhaps there is no better way to bolster the last argument than by showing how, throughout the history of premolecular genetics, the abstract nature of the genetical hierarchy has often been explicitly recognized. For instance, in 1915, even though the Morgan school had fully accepted that the chromosomes provided the "material basis of inheritance," they go on to observe:

> But it should not pass unnoticed that even if the chromosome theory be denied, there is no result dealt with in the following pages that may not be treated independently of the chromosomes; for, we have made no assumption concerning heredity that cannot be made abstractly without the chromosomes as bearers of the postulated hereditary factors. (Morgan et al. 1915, p. viii).

By 1925 the Morgan school had investigated about 400 mutant characters of Drosophila. These could be partitioned into four linkage groups corresponding to the four chromosome pairs. However, they observe:

> Were there no information as to the relation between the visible chromosomes and the linkage group it would still be possible to deal with the situation exactly in the same way by treating the linkage groups as a series of points held together in definite relations to each other. We might then speak of such groups as genetic [as opposed to physical] chromosomes.[49]

For investigative purposes, especially in the design of experiments, the additional knowledge that the linear order of the loci could be interpreted as a linear arrangement of positions along the chromosome was, of course, invaluable. As they put it:

> But since there is abundant evidence at present to show that the linkage groups correspond to chromosomes it would be only pedantic to treat the linkage groups without respect to the chromosomes, i.e., to treat heredity as a purely formal mathematical problem. Moreover, quite aside from this aspect of the subject, the known behavior of the chromosomes has at times furnished the key to genetic problems, and, conversely, the restriction of this situation to the known limitations of chromosome behavior has kept the subject from expanding into unprofitable fields of speculation.

Thus, though the coincidence of the two hierarchies was an empirical fact – and experimentally a very useful one – linkage relationships did not require, and could be consistently used without, such a coincidence. In 1926, on the basis of linkage data, Hamlett (1926, p. 442) even suggested a branched structure for the third linkage group. However, he immediately interpreted it chromosomally and suggested that the "resulting chromosome [was] thus of a type which [had] hitherto not been reported: that is, one in which the genes are not in a single linear series."

In the 1930s Haldane suggested that the physical order of the location of genes on chromosomes and the order of the loci inferred from linkage analysis would turn out to be inconsistent. This was going to show that, ultimately, mechanistic explanation would fail in biology.[50] Finally, in the 1950s, Lederberg et al. (1951) interpreted data from bacterial conjugation crosses, which is a type of linkage mapping in a prokaryotic context, to indicate a branched structure for the set of loci (see Figure 5.6.1). They were explicit that this structure was not to be interpreted as "a true branched chromosome." As they put it: "In a purely formalistic way, these data could be represented in terms of a 4-armed linkage group, Figure [5.6.1], without supposing for a moment that this must represent the physical situation" (p. 417). To classical geneticists, at least, the point that linkage analysis required an abstract hierarchy of alleles and loci in the genome, and not necessarily an identical physical hierarchy (though that may also be true), was usually obvious.

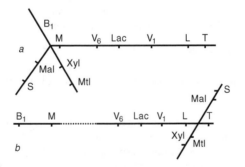

Figure 5.6.1. Linkage Data from *Escherichia coli* [after Lederberg et al. (1951), p. 417; by permission of Cold Spring Harbor Laboratory Press]. Note the branched structure of the map. In their caption, Lederberg et al. emphasize that "[t]his diagram is purely formal and does not imply a branched chromosome." Thus, linkage analysis establishes an abstract structure of the genome that is conceptually distinct from the physical structure.

But, how reliable are reductions based on this abstract hierarchy using techniques such as segregation or linkage analysis? If all the assumptions of a model used in the analysis (including the rules of gene transmission) are satisfied and in a statistical sense the fit is good, these inferences to a genetic explanation are as reliable as any in science. Unless one has an a priori philosophical prejudice against abstraction, in which case one eschews the probability of theoretical explanations in general, mere abstraction should not be taken as indicative of any special epistemological flaw of classical genetics. Nevertheless, the potentially confounding factors discussed in the last section demand that claims of genetic explanation should be treated with caution. This is particularly important because, as indicated in Chapter 1, genetic explanations or reductions have become socially fashionable, leading to the general acceptance of spurious explanations. The problems with the putative genetic explanation of manic depression have already been pointed out in § 5.4; the next section will discuss human male homosexual orientation and alcoholism. (The dubious status of heritability, as discussed in Chapter 4, disposes of IQ differences, religiosity, etc.) These criticisms do not argue against the possibility or the value of genetic reductions in general. Rather, they point out that for these reductions to be successful, a host of attendant assumptions must be met. Unfortunately this is precisely the point that has too often been ignored in the reductionist claims noted in the first chapter.

The most important reasons for caution in inferences to genetic reductions will be reiterated here. First, note that segregation and linkage analyses are both statistical techniques and, in complicated pedigrees, whether the models fit the data becomes a question of statistical significance. Spurious results can be obtained if not all the assumptions of the models used are met, for instance, if mutations are taking place when the models used do not permit that possibility. Even more important, the confounding factors discussed in § 5.5 can lead to spurious genetic explanations.

The critical problems are with the existence of phenocopies and variable expressivity. The existence of phenocopies may confound the statistical inference, as has already been mentioned. But they raise a more important interpretive problem. Strictly speaking, the segregation or linkage analysis of a pedigree can at most point to a genetic explanation of some trait in that pedigree. However, the usual question is whether that trait can be so explained in general. Sometimes this is not possible, which is why some phenotypes eventually get subdivided into smaller groups on the basis of what can and cannot be explained from a genetic basis. Thus, there are familial forms of diseases, which may admit genetic explanation, while the disease as characterized at the purely phenotypic level does not admit such an explanation in general. This distinction is harmless as long as there is an explicit delineation of what can be explained from the genetic basis. Unfortunately, enthusiasm for genetic reductionism often results in this distinction being ignored.

Variable expressivity is ignored by both pedigree and linkage analysis. However, whenever it is present, there is an indication that the explanation of the occurrence of the trait and its properties requires more than the loci that were investigated. It may require alleles at other loci or environmental factors. There is no systematic way of dealing with variable expressivity in general. It should come as a warning that even when a locus has some role in explaining the properties of a trait, that role may be very small. In such a situation, if many other loci are also involved in the explanation of a trait, they may still play such a significant role together, and one could still be justified in saying that a genetic explanation is obtained. Whether this ever happens is an empirical question; so far, no one has presented such a case. It is reasonable in this context to assume that genetic explanations are forthcoming only if variable expressivity is not found. In other words, genetic explanations are likely to be forthcoming only if techniques such as segregation and linkage analyses point to single loci with fairly constant and high expressivity.[51]

5.7.   NEW DIRECTIONS

Because they generally assume very simple models of the genotype-phenotype relation, segregation analysis and traditional linkage analysis permit inference to a genetic explanation only if one (or a very few) loci are sufficient for explanatory purposes. It has been clear since the beginning of genetics (and even earlier, in a sense, if one keeps the history of biometry in mind) that many traits cannot be explained by genes in this way. Nevertheless, genes may influence them and, since the 1980s, many techniques have been developed primarily to attempt to find such genes. Usually it is assumed in these cases that each locus has a small influence and that many loci are involved in the phenogenesis of a trait. One consequence of this situation is that, for any allele at a single locus, variable expressivity is likely to be diagnosed because of the potential allelic heterogeneity (that is, variation in the alleles) at the other loci that are involved. If this variability in expressivity is extreme, the problem of incomplete penetrance can also arise because for some combination of the alleles at other loci, the trait may not be exhibited at all.

Three of the most interesting of these new strategies for detecting genetic influence will now be briefly described. How they will fare in the long run is open to question.[52] Note, however, when there is variable expressivity, it is far from clear that alleles at such (or for that matter at all) loci alone, that is, with no reference to nongenetic factors, can ever provide the best explanation for a trait. Therefore, these new strategies may well be irrelevant for genetic reductionism for the reasons mentioned at the end of the last section. Unfortunately, this point is generally ignored by the advocates of these new strategies.[53]

(i) *Allele-Sharing Method*: Of the new strategies that have been developed, the allele-sharing method is the closest to traditional segregation analysis. However, like linkage analysis, it relies on the failure of one of Mendel's laws rather than its applicability. Unlike linkage analysis, it searches for the failure of the law of segregation of alleles. The basic strategy is to show that two relatives (usually siblings or cousins or uncle-nephew pairs in human populations) with the same trait inherit identical alleles at some locus with a frequency greater than what would be expected from the law of segregation of alleles if that allele had no role in the genesis of the trait.

For the purpose of genetic explanation, allele-sharing studies have two advantages over conventional segregation analysis: (a) they are more reliable for phenotypes with low penetrance. Thus, they can be used to detect alleles that have a small role in the genesis of these phenotypes. The reason for this is simply that, since only individuals exhibiting a trait are considered, low penetrance would only hurt the prospect of genetic explanation. It could not introduce spurious explanation. And (b), allele-sharing studies are relatively unaffected by the presence of phenocopies. This is an empirical point: since only close relatives are usually considered, it turns out that it is unlikely that phenocopies exist in such a fashion that spurious genetic explanations are introduced. However, there is one problem with this method that does not affect the traditional methods: it does not distinguish alleles that are identical by state but not by descent (for instance, through a new mutation) from those that are identical by descent.[54] However, this possibility is probably rare: to infer identity by descent from identity by state is usually reasonable.

Perhaps the most controversial application of this technique has been to a type of male homosexual orientation. The original study reported that brothers who both exhibited such an orientation showed excess allele-sharing (thirty-three out of forty pairs) at the locus Xq28 (on the X chromosome). A follow-up study reported the same result, but with only twenty-two of thirty-two pairs now showing allele-sharing. It also reported no such linkage for female homosexuality. However, in both cases there was an important caveat: while each pair of brothers was shown to share the same DNA sequence at the Xq38 location, different

pairs did not necessarily have the same sequence.[55] Attempts to reproduce this result have so far not been uniformly successful. As noted,
the follow-up study provided confirmation, but with a lower degree
of statistical significance. Another group failed to reproduce the result
altogether.[56] Moreover, methodological problems – including, possibly, the biased reporting of data – have been alleged in the work of the
original group.[57]

(ii) *Allelic Association Studies*: The principle behind these is simple. They
simply involve an attempt to find statistically significant correlations
between specific alleles and traits. Thus, they do not rely on any particular mode of inheritance. The general problem of moving from statistical
correlation to explanation manifests itself in two particularly important
ways in the interpretation of allelic associations: (a) positive correlations will be found between a trait and an allele, even if that allele has
no role in the genesis of the trait, provided that it is in "linkage dis-
equilibrium" with some allele that does play such a role. One standard
way in which this can happen is when both alleles are specified by
the same chromosome. In such a situation, two alleles at different loci
can occur together more frequently than expected for a variety of reasons, including genetic drift in small populations, population mixing or
migration, or genetic hitchhiking; (b) more importantly, positive correlations can be an artifact of population structure, especially population
admixture. In mixed populations, if some trait occurs with a higher than
average frequency in a subpopulation, there will be a positive association between it and any allele that also occurs with higher than average
frequency in that subpopulation.

The last point is well-illustrated by the case of the alleged association between alcoholism and the dopamine D2 receptor (in humans). It
had long been suggested that alcoholism is familial in the sense that it
is more likely in families with at least one alcoholic than in the general population (Cloninger 1987). The original report associating the
dopamine D2 receptor with alcoholism compared postmortem samples
from thirty-five alcoholics and thirty-five controls, without controlling
for any population characteristic except race.[58] The A1 allele was found
in 69 percent of the alcoholics and 27 percent of the controls. Subsequent attempts to replicate this result have yielded conflicting results,
with no association found in many cases and only an association with
severe alcoholism in others. A meta-analysis considered four reports
that claimed statistically significant associations and four that denied
them. It came out cautiously in favor of an association.[59] However,
what is critical is that in homogeneous populations, association studies

report no correlation; it is quite likely that the initial report and the other positive reports are simply artifacts of population admixture.[60]

(iii) *Quantitative Trait Locus (QTL) Mapping*: This method is closest to traditional linkage mapping. For populations that can be easily experimentally manipulated (that is, nonhuman populations, especially those with short generation times), Lander and Botstein's (1989) QTL mapping technique potentially allows the identification of the alleles at many different loci that can play a role in the origin of polygenic and quantitative traits. Two pure lines (say *A* and *B*) are crossed to form the next (F1) generation. Each individual then has one chromosome from each parent. However, crossing-over (leading to recombination between loci) can occur between them. The F1 generation is then back-crossed to one of the original lines (say *A*). In the F2 generation individuals will only be heterozygous with respect to alleles from the *B* line. The effects of these alleles can now be studied by searching for alleles known to differ between the two lines. If any of these correlate with a phenotypic difference (even in degree), the conclusion is that the corresponding loci play some role in the origin of that phenotype.

Going beyond what was said at the end of § 5.6, a final cautionary note is in order when genetic explanation is attempted using these new techniques. When many different loci are supposed to be involved in the genesis of a trait, and these techniques have to be invoked, there are two possibilities: (i) the trait is controlled at many loci, at least some of which have low expressivity (that is, the part of the trait that each locus controls has low expressivity). In such a case, it is unlikely that genes alone provide the best explanation for the trait, that is, that there is a genetic reduction; and (ii) the trait is controlled at many loci and each locus has high expressivity (in the sense given above). In this case there is still a potential for genetic reduction.[61] However, what must be remembered is that these techniques have so far only been used to identify individual (or a very few) loci that have some role in the origin of a trait. Therefore, so far no situation of the sort envisioned in (ii) has been found. In that sense, no genetic reduction with these new techniques has so far been obtained. Geneticists acknowledge this point implicitly by using the locution that these methods (and many others) only show that a trait has a "genetic component" that is quite different from simply claiming that it is "genetic." They also usually note, if not emphasize, that genes alone do not explain the phenogenesis of these traits. Unfortunately, the recent tendency to use genetic language at the expense of everything else, especially in semipopular and popular accounts of this work (as noted at the beginning of this book) lead to these nuances being lost.

6

Reduction and molecular biology

The types of reduction in genetics that have so far been considered in this book have been (a) and (c), that is, weak reduction and abstract hierarchical reduction, respectively. The former, considered in Chapter 4, assumed no structure at all for the genotype. The latter, considered in Chapter 5, assumed the genotype to be hierarchically organized, but did not assume that this hierarchy (of linkage groups, loci, and alleles) is one in physical space. This chapter will turn to strong reduction [type (e), also called "physical reduction," see below], which is the type most conventionally associated with the so-called molecular revolution in biology. This type of reduction assumes a spatial basis for a hierarchical reduction, that is, quite literally, the behavior of wholes is supposed to be explained by those of their constituent parts. In this chapter, unless explicitly indicated otherwise, "reduction" will only be used here in the sense of "strong reduction."

In genetics, given that the physical objects associated with genes are microscopic parts of cells, strong reduction is almost necessarily a reduction to an **F**-realm that, to some level of approximation, must be one in which the basic interactions are physical or chemical. It is tempting, therefore, to take the theory that is supposed to describe all interactions of matter at this level of organization, quantum mechanics, and its domain as the **F**-realm for these reductions. However, it will turn out that this will make reductions impossible because of the number of epistemically questionable approximations that have to be accepted on faith (see § 6.2 and § 6.3). (Moving to a slightly higher level of organization, that of quantum chemistry, does not turn out to be of much help.) Nevertheless, strong reductions in molecular biology have been remarkably successful, provided that the **F**-realm is taken to be what is perhaps best called the realm of macromolecular physics. Strong reductions in molecular biology will also be called "physical reductions," and the thesis that these can (or are at least likely to) be successful will be called "physical reductionism." This terminology has the fault of being misleading insofar as it suggests a reduction to physics at lower levels of organization than macromolecules (which are the levels at which physics is most commonly studied). However, keeping this explicit admonition in mind, this remains the most convenient terminology that can be adopted at this point.

The power of physical reductionism in molecular biology will be illustrated using five characteristic examples in § 6.2. However, before that, some general remarks on the role of reduction (and structural explanation) in the development and contemporary practice of molecular biology will be made (§ 6.1). The role of approximations in these reductions – which will indicate why macromolecular physics, rather than physics at some lower level of organization, must be taken as the F-realm – will be explored in § 6.3. Explicit attention to genetics will begin in § 6.4. Section 6.5 will briefly summarize the well-known complexity of giving a molecular "definition" of the gene. Section 6.6 will take up a question that has been a source of continuing controversy among philosophers, whether Mendelian (or classical) genetics has been – or is being – reduced to molecular genetics. In general, it will come out in favor of reduction. However, using the case of dominance as an example, § 6.7 will point out how strong reduction in molecular biology may fail. Finally, § 6.8 will emphasize the importance of not conflating physical reductionism with genetic reductionism, the subject of the previous chapter.

6.1. THE MOLECULARIZATION OF BIOLOGY

Physical reductionism in biology, traditionally called "mechanism,"[1] has a continuous history dating back to the mid-nineteenth century. It was vigorously debated during the first three decades of the twentieth century, generally pitting physiologists such as J. S. Haldane on the antireductionist side against the proponents of the new biochemistry, led by F. G. Hopkins.[2] The physiologists argued that irreducibly systemic properties of physiological complexes are necessary for the explanation of important biological phenomena, such as homeostasis or the Bohr effect, which show cooperative interactions between macromolecular parts (see § 6.2 below). The mechanists promised that ongoing developments in biochemistry, which provided a more "dynamic" theoretical framework than traditional organic chemistry, would ultimately reduce all such phenomena to ordinary chemical interactions.[3]

Nevertheless, at least arguably, successful reductions of this sort emerged not quite so directly from the work of the biochemists as from the attempts of Pauling and his collaborators to reduce chemical bonding theory to (the then newly discovered) quantum mechanics in the late 1920s and early 1930s. This resulted in a "structural" orientation that characterized many of the

major research programs that contributed to the establishment of molecular biology.[4] Pauling popularized the technique of constructing models of
macromolecules using atomic model kits where the atoms of different elements are simply taken to be spheres of different radii with appropriate bonding properties (including valency and the ability to form hydrogen bonds).
One of the most surprising facts about molecular biology has been the empirical success of these models (see § 6.2 below). An important early success
was Pauling's 1950 prediction that the α-helix would be a standard structure
to be found in proteins.[5] This prediction began to be confirmed almost immediately by Perutz (1951) and his collaborators who, drawing on the earlier
work of Bernal (e.g., 1939), had begun developing the crystallographic techniques necessary for solving the structure of large macromolecules. By the
mid-1960s details of the structures for several macromolecules, including
myoglobin and hemoglobin, began to become available, providing a means
to test various propositions such as "structure determines behavior" that had
become central to the theoretical framework of molecular biology. Starting
with the pioneering work of Levinthal (1966) and his colleagues, these two
approaches were merged through the development of computer modeling of
biological macromolecules.

Perhaps equally important, at least to set the stage for molecular genetics,
was the elucidation of the physical basis for the genetical hierarchy discussed in the previous chapter (§ 5.6). This began with Boveri's (1902) and
Sutton's (1903) discovery that the chromosomes were the hereditary material, continued with the Morgan school's establishment of the linear order
of loci on Drosophila chromosomes, Muller's (1927) important discovery
that radiations can induce mutations, and Painter's (1933) work on the physical structure of chromosomes. Shortly thereafter, Beadle and Tatum (1941)
convincingly demonstrated that individual genes were somehow responsible for the production of individual enzymes, thereby illuminating the mode
by which genes acted in organisms. The identification of DNA (rather than
protein) as the primary genetic material by Avery, MacLeod, and McCarty
(1944) and Hershey and Chase (1952) set the stage for the explosive growth
of molecular biology in the 1950s.[6]

Immediately after the definitive establishment of DNA as the genetic
material, making very effective use of Pauling's model-building strategy,
and drawing on the crystallographic work of Franklin and Wilkins, Watson
and Crick (1953a,b) proposed the double-helix model of DNA.[7] Its structure revealed a curious complementarity between DNA bases: adenine was
always paired with thymine (**A:T**), and cytosine was always paired with
guanine (**C:G**). Base pairing immediately suggested a mechanism of replication where each helix served as the template for the formation of another

"daughter" double helix. Such "semiconservative" replication was eventually experimentally demonstrated by Meselson and Stahl (1958). (Details of this mechanism will be discussed in § 6.6.) Base pairing also suggested a symbolic (rather than physical or chemical) interpretation and characterization of the relation between the DNA helices and, possibly, between DNA and other molecules in the cell. This played a role in the emergence of "information" as a concept to be used in molecular biology. The DNA sequence symbolically "encoded" all the hereditary information. Gene expression was a "translation" of this information; gene replication was its transfer to the next generation.[8]

Structural studies were by no means the only routes to molecular biology.[9] Other strategies from classical genetics, including the use of transmission genetics, though now pursued at the level of microorganisms, provided many of the problems and techniques for molecular biology in the 1950s. It brought with it the type of reasoning [including some explanations that are only abstract hierarchical reductions of type (c)] that was analyzed in Chapter 5.[10] Meanwhile, as noted above, "information" emerged as a potentially unifying concept that would provide molecular biology with its own distinctive conceptual framework (different from those of physics or chemistry).[11] Explanations in molecular biology involving the use of "information" – for instance, in the attempts to account for (and also predict) the properties of the genetic "code" in the 1950s, before it was experimentally deciphered – are (at best) approximate hierarchical reductions [of type (b)]. "Information" was involved in such theoretical claims as the central dogma of molecular biology, that information flows from nucleic acids to proteins, but never in the reverse direction (Crick 1958). Much of this theoretical framework proved to be sterile even in the 1950s and 1960s, when it was at least plausible, and collapsed in the 1970s as the eukaryotic gene turned out to have a much more complex physical representation (at the level of DNA) than ever expected (see § 6.5 below).

Nevertheless, information-based precepts do provide some useful explanations in molecular biology. For instance, they help explain why the code is nonoverlapping, or why DNA segments "encoding" alleles rarely overlap. If the code were overlapping, a single point mutation could modify several amino acid residues of a protein. If loci overlapped, a single point mutation could modify several alleles. Both of these possibilities, if realized, would decrease the stability of the information content of the genome. Therefore, the argument goes, these possibilities have been very strongly selected against during the course of precellular (or perhaps unicellular) evolution. Information, as connected to selection in this way, does seem to provide some insight into the biology at the molecular level.

To a much greater extent than information-oriented thinking, the modes of explanation and experimental strategies inherited from classical genetics have continued to be useful in molecular biology, resulting in the explosive growth of molecular genetics. In particular, they led to the formulation of the operon model of (prokaryotic) gene regulation, arguably one of the most important successes of molecular biology.[12] Though Monod (1971) later interpreted the operon as a cybernetic system, it provides a superb example of strong reduction, as was noted in Chapter 3 (§ 3.6).[13] Whatever plausibility the cybernetic framework might have had in the 1960s, it vanished in the 1970s, once again because of what Watson, Tooze, and Kurtz (1983) have very aptly called the "unexpected complexity of eukaryotic genes." Attempts to extend the operon model to eukaryotic gene regulation have been a dismal failure.[14]

Nevertheless, the most successful explanations in molecular biology have continued to be those in which structural considerations have been paramount, that is, physical reductions.[15] The success of Pauling's α-helix has already been mentioned. Pauling and his collaborators were equally successful in predicting yet another type of protein structure (the β-sheet) from their atomic kit models.[16] An equally spectacular success was the model of allosteric interactions constructed by Jacob, Monod, and their collaborators to explain cooperative behavior of proteins.[17] This example will be discussed in some detail in the next section (§ 6.2). It is important to emphasize that in these explanations, even leaving aside those that invoke "information," the assumptions are not codified into a single theory. The result of this situation is that molecular biology does not have a single unifying theoretical framework, though there are fragmentary theories, and much reasoning that can only be considered theoretical.[18] In § 6.3 an attempt will be made to state explicitly the main theoretical assumptions involved in the strong reductions of molecular biology. (No attempt will be made here to codify other frameworks that are used, for instance, information-based reasoning.[19])

At least up to the cellular level, the attempts to explain biological behavior, characterized at the molecular level, on the basis of the structure of macromolecules has been remarkably successful. (Some examples will be given in § 6.2.) One consequence of this has been the emergence of physical reductionism as the dominant ethos of much of contemporary biology: the pursuit of (strong or physical) reduction (that is, explanation at the molecular level) has come to dominate biological research programs in all areas of biology.[20] Two critical technical developments; recombinant DNA technology in the early 1970s and, particularly, the polymerase chain reaction, which allows very rapid amplification of specific DNA sequences (Mullis et al. 1986), have spawned the use of molecular techniques in even those other areas where molecular characterization and explanation are not directly of interest. Rather, molecular characterization, especially in the form of amino acid residue or DNA sequences, is used for purposes such as phylogenetic reconstruction or to test alternative theories of evolution. Within molecular biology they have helped create a concentration of interest on DNA over other types of molecules; some of the issues this raises for reductionism will be analyzed in § 6.8.

Returning to the situation where explanation is in the immediate agenda, in general, the progress of molecular biology has revealed a qualitatively new level of complexity of biological phenomena, when characterized at the molecular level. The revelation of this complexity – describing and classifying the wealth of structures and processes (the "mechanisms," as they are usually called) – has often been the primary pursuit of molecular biologists. What should be emphasized, however, is that this descriptive task, for all its allure, is but a prelude to explanation. Explanation (and, therefore, reduction) will be the main focus of this chapter. As the discussion of § 6.3 will show, molecular explanations from structures proceeds on the basis of a few relatively straightforward principles governing the interactions. Because of this, and because structure becomes paramount in these contexts, the elucidation of structure, which is often nontrivial, sometimes appears to be all that there is to molecular biology. An attempt will be made throughout this chapter to emphasize how explanations proceed from structural details. However, it is fair to say (using the terminology of Chapter 1, § 1.3) that the explanatory weight in these explanations is often borne more by the structures than the interactions.

Even in those areas where molecular explanations clearly cannot provide the answers to the questions that are perceived to be most important, especially in ecology and evolutionary biology, molecular techniques – especially DNA-based techniques – have become remarkably popular. As noted before, molecular homology (and analogy) is now routinely used to reconstruct phylogenies. Environment-organism interactions are now routinely characterized through molecular interactions. These attempts to extend molecular biology into new areas can at least partly be viewed as a standard type of scientific strategy: use a mode of explanation or investigation that has been successful in one domain in others as long as it is possible. However, as critics of the "molecularization of biology" point out, the success of a mode of explanation in one domain does not guarantee its success elsewhere. At levels of organization higher than the cellular level, obvious successes of molecularization have been rare. In particular, in many areas – including ecology and evolutionary biology – the advent of molecular biology has often been little more than an identification of some molecules (allegedly "key" molecules) that may be involved in the relevant interactions. However, the detailed elucidation of structure, on which the classic, and uncontroversial, successes of molecular biology were based, followed by reasoning about structure, have generally been absent. Moreover, in some areas, such as developmental biology, attempts at molecularization have even led to the avoidance of problems that have traditionally been of interest.[21] There is, therefore, ample reason for skepticism that the continued incursions of molecular biology into these other areas will lead to greater insight.

6.2. EXAMPLES

Five examples, chosen to illustrate the variety of strong reductions in molecular biology, will be analyzed in this section. All of them illustrate the success of the structural orientation mentioned in the last section. The first two involve the processes and interactions that are most often encountered in molecular biology. The third, allostery, shows how one of the standard types of example that traditional holists used as evidence against reductionism, namely, cooperative behavior, can be given a typical reductionist explanation. Similarly, the fourth, the operon model of gene regulation, shows how another type of example from the holists' traditional repertoire, namely, feedback control, can also be brought under the reductionist rubric. This example was already discussed in detail in Chapter 3, § 3.6. It will suffice, here, only to note its significance to the present context. Finally, the fifth, the case of sickle cell hemoglobin, is one where reductionist explanation traverses from the phenotypic level of a disease to that of the DNA sequence.

(i) Immune recognition, a highly specific cellular and chemical interaction, is a critical part of the immune response of higher mammals. During that response, those B lymphocytes that have antibody receptors with the highest affinity for an antigen (this is what is called "recognition") proliferate more rapidly than other lymphocytes. This process, known as "clonal selection," results in a huge population of B cells with very high affinity for that antigen. (In clonal selection, affinity plays the role of fitness in natural selection.) Subsequently, the structures bearing that antigen are chemically degraded and removed from the body. What mediates immune recognition is a very close lock-and-key fit between the antigen and a part of the antibody's surface. Both regions are generally hydrophobic [which provides an entropic force driving their association; see (ii) below]. For instance, when lysozyme is the antigen, its binding to its antibody results in contact at a 600 $Å^2$ surface, involving sixteen lysozyme and seventeen antibody amino acid residues. A similar area of interaction is seen when neuraminidase is the antigen.[22] The four most important points to note about these lock-and-key fits are that: (a) complementary molecular shapes are paramount insofar as differences in the shapes of the interacting surfaces of the two macromolecules would destroy a fit; (b) at least approximately, all parts of the molecular surfaces that are in contact with each other are equally important; (c) the interactions seem generally to be driven by hydrophobicity and the formation of hydrogen bonds; and (d) the interactions are weak. These points will be taken up in some more detail at the end of § 6.3. Suffice it here to note that the immune recognition example is typical; similar lock-and-key fits provide the molecular mechanism for a very large majority of the phenomena that are explained by molecular biology.

(ii) By definition, hydrophobic substances are at most sparingly soluble in water (though they may be soluble in organic – that is, nonpolar – solvents).[23] Hydrophobic substances tend to clump together; this interaction is sometimes called the "hydrophobic bond."[24] The reason for this was for a long time incorrectly attributed to attraction between these substances though the correct mechanism for it had been partly recognized as early as the 1890s.[25] The structure assumed for (liquid) water plays a critical role in accounting for the hydrophobic effect. It is assumed that, in the absence of solute molecules, the water molecules are isotropically arranged and each water molecule forms four hydrogen bonds with others. This is a critical approximation and will be analyzed in detail in § 6.3. If a solute is present, this structure is disrupted. If the solute is polar (hydrophilic), its molecules also form hydrogen bonds with water molecules and structures similar to those that would occur in its absence are recovered. No such compensation occurs for hydrophobic "solutes." The result is the distortion of the hydrogen bonds of the water and the formation of more such bonds. A consequence of this is the increase of the (Gibbs) free energy (or more accurately, the decrease of entropy). If the hydrophobic substance is confined to the smallest possible volume, this increase is least. Consequently, hydrophobic substances tend to clump together in an aqueous medium. Thus, the hydrophobic effect is driven by the free energy (or more accurately, the entropy) change.

This picture is qualitative, but parts of it have been quantitatively modeled. What is more interesting in this context is that values of hydrophobicity, measured on a suitable relative scale as determined through experimental measurements, can be used to account for many facets of the behavior of biological structures, especially membranes (including protein–membrane interactions). In particular, amino acid residues in proteins can be given (average) values of hydrophobicity on the basis of hydrophobic/hydrophilic interactions of their side chains. Glycine with no side chain is used as the origin of the scale, with hydrophobicity 0. Hydrophilic residues have negative value for hydrophobicity, with the four charged residues (lysine, arginine, glutamic acid, and aspartic acid) having the most negative values; hydrophobic residues have a positive hydrophobicity. These values are extremely useful. They can, for instance, be used to explain how segments of some proteins such as colicin E1 enter into a bacterial membrane to form channels for ion flow.[26]

(iii) Some proteins (usually called "oligomers") such as hemoglobin, which consist of several subunits (or "protomers"), show cooperative binding with their substrates or other reactants. Hemoglobin shows this phenomenon in its binding to oxygen. The result is a sigmoid (S-shaped) binding curve. Myoglobin, which consists of a single chain, shows the expected simple binding curve with oxygen. As oxygen pressure or concentration increases, myoglobin associates with oxygen to about 30 percent of the saturation value before any significant association of oxygen to hemoglobin occurs. Yet, when hemoglobin begins to associate with oxygen, the curve rises steeply and intersects with that of myoglobin at about 90 percent of the saturation value. This is known as the Bohr effect. Clearly, there is cooperative behavior in the case of hemoglobin – binding some oxygen helps hemoglobin to bind more. Monod and Jacob (1961) called proteins of this sort "allosteric."

Cooperative phenomena have traditionally been part of the standard repertoire of antireductionists: the "whole" appears to be "more than the sum of the parts," the relation of the whole to the parts is "nonlinear," and so forth. One of the most important developments within molecular biology was Monod, Wyman, and Changeux's (e.g., 1965) elaboration of a straightforward reductionist (MWC) model for allostery.[27] This model describes allosteric proteins as consisting of distinct spatial parts and requires ordinary (F-justified) assumptions about the interactions of these parts.

Following the modern treatment of Cantor and Schimmel (1980b), the MWC model makes four assumptions:

(a) identical protomers occupy equivalent positions in the oligomeric protein (in the case of hemoglobin, there are four protomers);

(b) each protomer contains exactly one receptor site for the reactant (ligand) (for hemoglobin, the reactant is oxygen);

(c) the oligomer has at least two distinct conformations accessible to it – the affinity of the reactant to the receptor sites may be different in these two conformations (in the case of hemoglobin, there are exactly two such conformations, and the affinities are very different); and

(d) this affinity depends on the conformational state of the oligomer (and, therefore, the monomers) but not on the occupancy of neighboring sites.

From these assumptions, using standard chemical kinetics, it is trivial to derive the S-shaped curve for binding. If (d) were not valid then a hierarchical explanatory graph could not be constructed – it is critical toward ensuring a straightforward reduction. Finally, to show that the detailed assumptions can be justified, what must be done is to show that the active site-reactant interactions are governed by the standard ideas about lock-and-key fits that were described above.

(iv) The operon model was already discussed in detail in Chapter 3, § 3.6. Its significance, in the present context, lies in the fact that it explains regulation through feedback. β-galactosidase, which digests lactose, is only produced in the presence of that substrate. Feedback regulation, which used to be called "homeostasis" by physiologists, was also traditionally part of the antireductionists' repertoire. Systems exhibiting feedback were supposed to have such complex interactions between the parts that their (conceptual) dissociation into such parts that the interactions of which alone explained the behavior of the whole was supposed to be impossible. Explanations were supposed to have to refer irreducibly to states of the whole.[28] The operon model brings it squarely within the domain of what can be accounted for by strong reduction.

(v) Perhaps the case that best demonstrates the power and also the limitations of strong (physical) reduction in molecular biology is that of the sickle-cell trait. In 1910 Herrick (1910) examined an anemic patient and observed that his blood contained a large number of "sickle-shaped and crescent-shaped" red blood corpuscles. Over the next decade this observation was repeatedly confirmed and in 1922 Mason (1922) coined the name "sickle cell anemia" to describe the disease. As studies moved to the biochemical level, Hahn and Gillespie (1927) demonstrated that sickling was associated with the deoxygenation of hemoglobin. In 1934 Diggs and Ching (1934) showed that sickled cells have difficulty (unlike normal cells) in passing through the capillaries, thus establishing how the presence of these cells could lead to anemia.

Pauling began work on sickle cell disease in 1945. Using electrophoresis, he and his collaborators demonstrated that there was a difference in the number of ionizable groups between the hemoglobin of normal cells (HbA) and that of sickled cells (HbS).[29] They also showed that this difference must be one between the polypeptide chains of HbA and HbS (rather than in the heme groups). Individuals with sickle cell anemia had only HbS, those with the sickle cell trait but not anemia had both HbS and HbA in roughly similar amounts, and those who were normal (with respect to this trait) had only HbA. The conclusion was that HbS arose from a mutation at a single locus. Ingram (1956) showed that the difference between HbA and HbS was in only one amino acid residue, at position 6 of the β-chain of hemoglobin: whereas HbA had glutamic acid, HbS had valine instead.

This difference is revealing. In HbS a hydrophobic amino acid residue (valine) occurs instead of a charged and, therefore, highly hydrophilic residue (glutamic acid) in HbA. The concentration of hemoglobin in red blood cells is so high (340 mg/cm^3) that it is close to crystallization under normal circumstances. When a hydrophilic residue is replaced by a hydrophobic one, aggregation is even more likely, especially if the hydrophobic region finds a complementary (receptor) site on another molecule. Apparently such a site is present only in deoxy-HbS (and not in oxy-HbS).[30] Murayama (1966) proposed that the HbS molecules spontaneously formed thin helical fibers that led to the sickled shape of the cells. The details of this model turned out to be incorrect. At present, thicker helical fibers of either fourteen strands (proposed by Edelstein and his collaborators) or sixteen strands (proposed by Josephs and his collaborators) seem more likely.[31]

Either of these models explain how deoxy-HbS spontaneously ag-gregate to form helical fibers that distort the red blood cells, ultimately leading to the symptoms of sickle cell anemia. In some sense the molec-ular explanation of the trait is complete, though many details may still have to be filled in. What should not go unnoticed, however, is that in spite of this detailed molecular account there is no cure, nor even any sig-nificant successful therapeutic intervention, for the disease. Molecular understanding is no assurance for therapeutic intervention. Paul (1995) constructs a similar cautionary story for PKU: molecular medicine ex-ists more in scientific mythology than in the clinic.

6.3. APPROXIMATIONS AND THE PHYSICS OF MACROMOLECULES

The discussion in the previous section emphasized the versatility and power of strong reductions in molecular biology. Quite intentionally that discus-sion, which tried to be accurate in all scientific detail, was ambiguous about the **F**-realm to which these strong reductions occurred. Much of the philosophical discussion of the issue of reduction in molecular biology is also ambiguous on this point. There have been three common proposals: (i) that (classical) biology is being reduced to molecular biology; (ii) that biology is being reduced to chemistry (or biochemistry)[32]; and (iii) that biology is being reduced to physics, in particular to quantum mechanics (which is usually taken also to have reduced chemistry to physics).[33] At first sight, the first of these proposals has little content given that strong reductions in molecular biology do not invoke specific "rules of molecular biology." Rather, what these reductions seem to be invoking are rules inherited, in some fashion, from physics and chemistry. Nevertheless, the first proposal is the one that, after being made more precise, will turn out to be the most correct. The other two proposals do not have similar merit. This is seen by analyzing how the sort of explanations discussed in the previous chapter can (i) neither be explanations from a quantum-mechanical basis (ii) nor one from (usual) chemistry.

(i) The most important assumption is that the biological macromolecules have a definite shape determined by the outer (exposed) surfaces of the atoms on the surface of the macromolecules. The atoms are assumed to be spherical and solid enough to present problems of steric hindrance (if they are supposed to intersect).[34] There are different ways to compute the radii of these spheres. The most common approximation is one in which the radius is set to half the length of the covalent bond between two such atoms after this length is determined experimentally. Similarly, resort to experiment is also necessary to find the shape of macromolecules. The most straightforward method is X-ray crystallography, which provides electron density maps which, in turn, specify the "centers" of atoms assumed to be spherical.

These hard spherical atoms, reminiscent more of Dalton's atoms[35] than of modern quantum mechanics, are the basic units out of which the models of molecular biology are constructed, whether using the atomic kits of Pauling or Watson and Crick, or the computer images pioneered by Levinthal. But can this picture of the atom – and the attendant theories about hydrogen and other bonds, hydrophobic interactions, and so forth – be explained, or in any way justified, from quantum mechanics, allegedly the fundamental theory at the atomic level of organization? Before that question is addressed, two points should be kept in mind. (a) The assumption that atoms can be treated this way has strong experimental support. Therefore, at least from an empiricist point of view, it is a moot question whether the models can be justified from some underlying basis. (b) Even if these properties of the atoms can be derived from a quantum-mechanical basis, and thereby be reduced to it, this does not necessarily mean that the biological phenomena can also be reduced to that basis. If each of these individual reductions involve complex approximations, they may not together constitute a reduction; as noted in Chapter 3, § 3.4, the "goodness" of approximations is not always a transitive relation.

But does quantum mechanics allow such atoms? Leave aside the questions of whether the reduction is of anything stronger than type (i) – weak reductionism – and whether the reduction of ordinary chemical bonding theory to quantum mechanics is truly successful; these are issues that are beyond the concerns of this book. Even heuristically, what does quantum mechanics allow for atoms in macromolecules? At present the answer is unclear. Certainly there is no reason for assuming that the solid balls assumed in the models can be obtained in quantum mechanics. For atoms in ordinary (small) molecules, Bader (e.g., 1990) has provided a systematic account that admits the generation of most usual chemical properties including bonding. As should be expected, this is an approximate treatment. In this theory (as should also be expected) atomic surfaces are not quite as fixed as in the models used in molecular biology, though the charge density is at a (local) maximum on the surface. The most counterintuitive part of this theory is that it does not, in a straightforward way, generate the usual Lewis (electron-pair) model of covalent bonds.[36] Thus, there may even be a problem of consistency[37]; certainly, the epistemic conditions for explanation are far from met. Whatever strong reduction in molecular biology is, it is not a reduction to the quantum realm.

(ii) But could reduction be taking place to only a slightly higher level of organization, say a chemical level, where solid spherical atoms are assumed to exist? The trouble now is that many of the critical biological interactions are mediated not by the usual covalent or ionic bonds of chemistry. Covalent bonds do hold macromolecules together. But their explanatory relevance in the context of molecular biology is slight. Rather, the explanatory weight falls on very weak interactions especially hydrophobic interactions.

Now, recall the discussion of hydrophobicity in the previous section. The quantitative account, to the extent that there is one, relies on a model of water that is isotropic, with each water molecule forming exactly four hydrogen bonds. Not only does this model make counterfactual approximations, these are counterfactual in a rather striking way: taken seriously, this is a static model of water, in direct contrast to the standard kinetic picture of liquids. Moreover, the water molecules are supposed to have rigid bonds. Once rigidity is dropped, and kinetics allowed, one does not even know how to begin the type of entropic calculation on which a quantitative account of hydrophobicity is based. These approximations are incorrigible (at least in practice), their effects are inestimable, and they are context-dependent. Similarly, unlike ordinary ionic bonds, the ionic bonds that may be important in molecular biology are weak, because of the aqueous cellular medium, and the bond energies have to be determined experimentally or through calculations based on counterfactually approximative models of the structure of water. If explanations in molecular biology are supposed to be reductions to the **F**-rules of chemistry, no such reductions are known in practice. Indeed, someone with a penchant for polemic can seriously raise the question whether these weak interactions are even consistent with the rules of ordinary chemistry.

Nevertheless, the examples discussed in the pevious section should have shown that molecular biology does provide striking explanations that appear to be strong reductions. The solution to the problem is to recognize that these explanations invoke **F**-rules that are peculiar to molecular biology and that provide it with a theoretical framework of its own. For want of a better name, these **F**-rules will be called the **F**-rules of "macromolecular physics." Though they have not been systematically codified, some of them can be easily stated. For instance, the following four rules have been routinely invoked in this chapter.

(i) Weak interactions rule: the interactions that are critical in molecular explanations are very weak. For instance, the covalent bonds of ordinary chemistry have a bond strength of 90 kcal/mole [that is, it takes 90 kilocalories of energy to break 1 mole (6.022×10^{23}) of these bonds]. Ordinary ionic bonds have a strength of 80 kcal/mole. In the aqueous cellular environment, hydrogen bonds have a strength of 1 kcal/mole and ionic bonds have a strength of 4 kcal/mole.[38] Hydrophobic "bonds" have a strength of 1 kcal/mole.

(ii) Structure determines function:[39] the behavior of biological macromolecules can be explained from their structures as, for instance, can be determined from crystallographic studies. This principle continues to motivate crystallographic and many other programs for determining macromolecular structures.

(iii) Importance of molecular shape: these structures, in turn, can be characterized entirely by molecular size and, especially, shape, and some general properties (such as hydrophobicity) of the different regions of the surfaces. In general, exposed surfaces of the molecules, that is, those that are in contact with water molecules, will be hydrophilic. Hydrophobic moeities will be found either inside molecules or on those parts of molecules that are found inside membranes or other such water-excluding regions.

(iv) Lock-and-key fit for molecular interactions: molecules such as protein molecules forming larger structures or enzymes interacting with substrates interact when there is a lock-and-key fit between the two surfaces. There is no interaction when these fits are destroyed.

These rules may well be "coarse-grained," that is, they are the result of the "averaging" of many lower-level processes, and may also only be approximate. It is possible that, as the molecular explanation of biological phenomena gets more detailed, more "fine-grained" considerations, for instance, at the level of the $\mathbf{F}$-rules of ordinary chemistry, will replace them. Returning to the first example treated in the previous section, that of immune recognition, there is already some evidence that not all parts of the interacting surfaces are equally important: only four or five of the residues may be critical, though ten to seventeen of them may form the interacting region.[40] Something more than structure and lock-and-key fits may be necessary to explain these interactions. Should moves of this sort become routinely necessary, then the account of macromolecular physics, and the role it plays in strong reduction in molecular biology, that has been given in this chapter will require modification.

6.4. THE STATUS OF GENETICS

So far, almost nothing has been said of genetics. However, it is this case, the potential reduction of classical genetics to molecular genetics, that has dominated almost all the philosophical discussion of reduction in molecular biology.[41] In choosing this emphasis, philosophers have followed molecular biologists, who have also increasingly focused on genetics since the mid-1960s, so much so that it can be plausibly argued that molecular genetics has become the core of molecular biology, in contrast to the 1950s and early 1960s, when the study of proteins and other molecules was as central to the new field as the study of DNA. However, it should be emphasized that even if reductionism is problematic in molecular genetics, it does not follow that it is similarly problematic in the rest of molecular biology.

Given what has already been said about the power of (strong) reduction in molecular biology in general, it might come as a surprise to note that there is no philosophical consensus about the status of reductionism in molecular genetics. Most philosophers of biology, following Hull's (1972) early objections, have rejected the position that reduction captures the relation between classical and molecular genetics.[42] A different objection was made by Rosenberg (e.g., 1978). However, Schaffner (e.g., 1993a,b) has continued to argue in favor of the validity of his model of reduction in this case.[43] Ruse (1976) has argued in favor of Nagel's (e.g., 1961) model of reduction, not even requiring the (slight) liberalization of it that Schaffner (1967b) had introduced.[44] Waters (1990) admits that the attempted reductions remain sketchy but argues that they are forthcoming and ultimately will be cases of theory reduction. He does not, however, fully commit himself to the logical empiricists' construal of theories that Schaffner (1993a,b) continues to use. Wimsatt (1976b) argues for reduction but rejects the requirement that it be a relation between theories. Instead, he offers his own model. Sarkar (1989, 1991, 1992) also argues for reduction and, like Wimsatt, rejects the requirement that it must be construed as a relation between theories. However, he does not endorse Wimsatt's (1976b) formal model and leaves open the question of the form of explanation.

The same position will be advocated here. The various objections, therefore, will have to be addressed. However, before that is done, one point must be clarified. Most philosophical discussions of this question pose it as one of the reduction of classical genetics to molecular genetics. In other words, molecular genetics is supposed to provide the **F**-realm. This seems to be at variance with the discussion earlier in the previous section where the **F**-realm that was appropriate for reductions in molecular biology was identified to be that of macromolecular physics, which seems to be rather different from genetics. This difference is less significant than it initially appears. In molecular genetics, the same processes as those that were partly codified in § 6.4 remain the relevant ones. The critical difference is that DNA (and, in a few cases, RNA) are the macromolecules that encode genes. In this sense, those processes now have to be considered in the special circumstance where nucleic acids play an important role. In the rest of his chapter the **F**-realm will be referred to as "molecular genetics" in order to remain in agreement with other discussions. This should generate no new confusion.

Following the spirit of this book, when analyzing the objections to reductionism in molecular genetics, the objections that depend on formal considerations will be distinguished from those that depend on substantive questions. Detailed answers will only be given to the latter. There are five objections that can be profitably distinguished.[45] The first four of these can all be traced back to Hull's (1972) paper where they are presented together (that is, not explicitly distinguished) as part of his general antireductionist argument. The fifth is most clearly articulated by Rosenberg (1985).

(i) The "no theory" objection: as noted in § 6.1, molecular biology has no central unifying theory. Moreover, it is generally conceded that molecular genetics fares no better in this respect. Hull (e.g., 1972), Wimsatt (1976b), Kitcher (1984), and Burian (1996) find no theory at all.[46] Sarkar (1989, 1992) only finds fragments of theories. Waters (1990) thinks that theories will eventually emerge, but they are not yet present. Only Schaffner (1993a,b) finds genuine theories. If reduction must be a relation between theories – as Schaffner (1993a,b) and many others have continued to advocate (see Chapter 2, § 2.3) – then unless Schaffner is right about there being theories in molecular genetics, the position that molecular genetics does now reduce classical genetics cannot be maintained. Moreover, Schaffner does not present any good candidate for a theory. His examples are even less general than the principles outlined in § 6.4 as being central to macromolecular physics.

This objection is of doubtful value. Almost all the disputants accept that molecular genetics significantly explains classical genetics.[47] [A possible "explanatory inadequacy" objection will be addressed separately later in this section as objection (v).] They also usually concede that molecular genetics forms a realm different from that of classical genetics, and that explanation – when available – proceeds by constructing hierarchical spatial models of the system. The dispute then becomes one about either the explication of "theory" or about the role of theories in reduction.[48] These, in turn, have been almost universally construed as formal questions and are, therefore, not particularly relevant to the discussions of reduction in this book. There are two possible conclusions to be drawn from this situation: (a) if reduction is necessarily a relation between fully articulated available theories, and if theories are to be construed in any way similar to those of the logical empiricists, then there is no current (or probably even ongoing) reduction of classical to molecular genetics; (b) if the substantive criteria about reduction are the important ones, then the emphasis on theories is misplaced in this debate. The discussions of this book are obviously far more concordant with the second of these possibilities. However, the argument given here does not allow that to be drawn as a logical conclusion.

(ii) The "different domains" objection: Hull (1972) argued that classical genetics is a theory of gene transmission, whereas molecular genetics is a theory of both gene transmission and development.[49] From Hull's point of view, for reduction to take place, both theories should have the same domain. This objection, as stated, is clearly based on an erroneous assumption about the nature of reduction; in general, reducing theories have larger domains than the reduced theory. However, there is an objection of this sort that has more merit: that classical genetics deals only with transmission (and reproduction) whereas molecular genetics only deals with gene expression (and development). Thus, to use the terminology of Waters (1990), the two domains are "unconnectable" through reduction (or for that matter any similar relation between the domains).

However, both these assumptions about the domain of classical genetics and that of molecular genetics are erroneous. Classical genetics, through its elaborate framework for the varieties of dominance, expressivity, penetrance, and in other ways (see Chapter 5), did attempt to deal with gene expression and the phenogenesis of traits. What is fair to say is that the classical account of expression (and, in general, of developmental processes) was far from adequate and often involved little more than genetic redescription of the phenotypic phenomena. It is also fair to say that molecular genetics, so far, only provides hints of an account of development even though it has elucidated fairly detailed models of individual gene expression. In particular, it does not yet provide an account of development; this point will be discussed in some detail in § 6.7. There are two philosophical conclusions to be drawn from this situation: (a) reduction of classical genetics to molecular genetics is far from complete; and (b) reduction proceeds in a piecemeal fashion, which is far from what theory reductionists such as Nagel (1961) or Schaffner (1967b, 1993b) envision as the process of reduction.[50]

Moreover, molecular genetics does attempt to explain gene transmission. It has a fairly detailed understanding of gene replication, the first stage of transmission. Moreover, the molecular explanation of the independent assortment of alleles on separate chromosomes is relatively trivial. Through the elaboration of mechanisms of recombination, explanations of linkage relationships are also emerging. More detail on these topics will be presented in § 6.6. While neither all of transmission genetics nor much about the phenogenesis of traits can yet be accounted for from molecular biology at present, in principle unconnectability is not a justifiable claim.

(iii) The "multiple realizability" objection: Both Hull [starting with his (1972)] and Rosenberg (1978, 1985) have pointed out that some phenomena in classical genetics can be the result of different mechanisms at the molecular level. In the context of the relation between mental and physical states, this objection has been called "multiple realizability." Kincaid (1990) has explicitly invoked this objection and this terminology in the present context.[51] It is less than compelling, unless it is coupled with some ontological requirement about the identity of states, etc., and some formal requirement about the form of the connections. These requirements were already discussed and rejected in Chapter 2 (§ 2.5). Consider the relation between macroscopic and microscopic physics: it is the standard situation that the same macroscopic state corresponds to a virtually infinite number of microscopic states. This is exactly the situation exploited by statistical mechanics to provide explanations (that are usually accepted to be reductionist). The multiple realizability objection would suggest that these explanations do not explain! In comparison, the number of mechanisms known to be responsible for most genetic phenomena (for instance, DNA repair or mitosis or meiosis) are remarkably few. Even if multiple realizability were a compelling objection in general, it would fail in the genetic context because of the remarkable homogeneity of the mechanisms and, perhaps even more so, because of the simple rules that govern macromolecular physics (see § 6.3).

(iv) The "no natural kinds" objection: Implicit in both Hull's (1972, 1974) and Rosenberg's (1978, 1985) discussions is a disquiet over the fact that there is no one-to-one correspondence between processes at the classical and the molecular levels. This, to use Fodor's (e.g., 1974) terminology, would lead to the situation where "natural kinds" in one realm did not correspond to natural kinds in another. The trouble with such an objection is that it replaces a possibly complex relation – for instance, that between classical genes and DNA (see § 6.5 below) – with an even murkier relation, that of identities between natural kinds. In the absence of any clear criteria for natural kinds either in classical genetics or, especially, in molecular biology, it is hard to judge whether this objection can even be clearly articulated. Moreover, in practice, worries about natural kinds rarely play any role in scientific work; at best these are only produced during philosophical reconstructions of dead disciplines. This objection will simply be ignored here with the suggestion that philosophical and conceptual analysis should remain more closely tied to the actual theoretical and explanatory context of the science being discussed rather than introduce even murkier notions that only seem to generate confusion.

(v) The "explanatory inadequacy" objection: Rosenberg (1978, 1985, 1994) has argued that classical genetics is only supervenient on molecular genetics. This objection can be taken as an extreme form of the last two objections. As already discussed in Chapter 2, § 2.6, Rosenberg (1978) maintains that the complexity of molecular accounts prevents any sensible explanations of classical genetics to emanate from them. This complexity is due to the multiple realizability of classical entities and processes, compounded by a failure to find (lawlike) connections between natural kinds. Rosenberg concludes that the relation between the two domains is one of supervenience. There are at least two responses to this objection.

(a) As Waters (1990) has emphasized, almost all disputants about reductionism in this context admit that molecular explanation is taking place. The dispute usually is about the form of these explanations. While this point does not refute Rosenberg – after all, all the other disputants could be wrong – it does suggest that his position may be implausible. Biologists, moreover, freely admit that molecular genetics provides explanations for phenomena of classical genetics.[52] Now, if reduction is ultimately no more than an appropriate type of explanation, as has been advocated throughout this book, that is, if ontological considerations are not critical, no more than the point noted in the previous section is needed in response to Rosenberg.

(b) More importantly, while molecular biology does provide a variety of mechanisms, these do not appear to be so diverse as to suggest that their invocation makes explanations incomprehensible. Though Rosenberg is correct in noting that molecular accounts are often startlingly complex, this complexity is not quite so inchoate as not to allow patterns to emerge. To some extent Rosenberg appears to be misled by the fact that the importance of the details of the mechanisms, especially structural details, is such that they, rather than the details of the interactions, bear more of the explanatory weight in molecular explanations.[53] It is Rosenberg's responsibility to provide even one example where molecular complexity is such that the mechanisms, when presented, do not explain because of their complexity, rather than because of the incompleteness of the present knowledge of the situation. As already mentioned once before, supervenience, in the present situation, seems to be the counsel of unnecessary despair.

Once again, what can be said with justice is that the molecular explanation of many phenomena is incomplete. However, explanatory incompleteness is characteristic of any living scientific discipline. It should only be interpreted as indicative of potential explanatory inadequacy if either inexplicable phenomena abound (that is, there are many clear anomalies) or if the explanatory project seems to generate no new insights. Neither is the case in molecular genetics, as will be shown in § 6.6. One of the few obvious failures to explain

a well-characterized classical fact, that of dominance, will be discussed in
§ 6.7.

6.5.   THE MOLECULAR "DEFINITION" OF A GENE

Ultimately, the worry that lies behind the last three objections addressed
in § 6.4 is the unexpected difficulty of defining the "gene" in molecular
terms. Though the antireductionists almost always state these objections using molecular genetics as it stood in the late 1960s, when only the prokaryotic
genome (especially that of *Escherichia coli*) had been explored, there is no
serious difficulty of definition in that context. Each segment of DNA was
associated with exactly one locus; alternative sequences were alternative
alleles; and all loci were either structural (that is, capable of specifying an
RNA sequence) or regulatory. The molecular definition of a classical allele
is, therefore, trivial ("a sequence at a locus"); and the definition of a classical
locus would be a particular segment of DNA, defined by its position relative
to other segments of DNA. Loci would be of two types: RNA-specifying and
regulatory. There was no known rule for determining boundaries between
loci at the DNA or RNA level. However, these loci could be systematically
located at least by enumeration. It takes more philosophical imagination than
insight to find these definitions problematic.[54]

Once these definitions are accepted, the question becomes whether rules
at the level of molecular genetics (the **F**-rules) explain the facts of classical genetics. Given that the question at this stage is confined to prokaryotic
genetics, where reproduction is asexual (and nothing even as sophisticated
as Mendel's laws has to be invoked), there is little reason for skepticism
about such explanation. As a historical point, however, it is at least tenden-
tious, if not misleading, to talk of the reduction of classical to molecular
genetics in this context. The prokaryotic analogs of the rules for eukaryotic
classical genetics were generally investigated only at the molecular level,
starting, roughly, in the late 1930s. There was never a well-articulated classical prokaryotic genetics independent of molecular explanations, so the
question of molecular explanation or reduction is relatively uninteresting.

However, in the eukaryotic context the developments that have occurred
in molecular genetics since 1970 can only please the antireductionists. In
eukaryotic genomes the representation of genes at the molecular level has
turned out to be far more complex than had even been envisioned during
the 1960s. The only significant complexity that was fully recognized before
1970 was that some genes were "movable." As early as 1947 McClintock
had observed them in maize: these are bits of DNA that can not only change

positions inside a chromosome or between chromosomes of an individual but can even move between individuals and between species.[55] These present no serious difficulty for the definition of an allele, but they already complicate the definition of a locus. In any individual the locus is still well-defined by its relative position in a particular chromosome. However, this definition is far from fully satisfactory: it could vary from individual to individual (within a species).

Developments since 1970 have been strange enough to make this seminal discovery look trivial as far as the complications of eukaryotic heredity are concerned. Five of the most interesting ones will be discussed here. The first four of these complicate the molecular definition of a gene. The last is mentioned here only to give a relatively complete picture of the complexity of modern molecular genetics.[56]

(i) Much of the DNA in eukaryotic genomes does not have any coding (structural) or regulatory role. This so-called "junk" DNA is estimated to constitute 95 percent of the human genome. It occurs both between regions corresponding to classical loci and within them. In the latter case it forms "introns." After transcription, portions of the RNA corresponding to the introns are "spliced" out. Moreover, alternative splicing (the production of different RNA segments for translation from the same original transcript) has also been found.[57] Thus, no straightforward identification of DNA with locus or DNA sequence with allele is possible.

(ii) Frameshift mutations have been discovered and have destroyed any residual belief in what used to be called a "natural synchronization" of the genetic code. In these, the DNA sequence is not "read" through, triplet by triplet, beginning with a particular nucleotide; there is a shift by one or two bases while "reading," leading to an entirely different amino acid residue chain. Though a few examples are known, the extent to which frame shifts are present in organisms is at present largely a matter of conjecture (Atkins et al. 1991). Sometimes, frame shifts are used to begin a segment of DNA that codes for a different protein (Fox 1987). Thus, sometimes different proteins are produced from the same DNA segment; these are types of "overlapping genes."[58] Thus, the relationship of DNA segments to loci may be one-many.

(iii) Besides splicing, several types of post-transcriptional mRNA modifications generally called "editing" are also now known.[59] DNA segments producing transcripts that are subsequently so edited are sometimes called "cryptic genes." For instance, in the intestines of mammals, the mRNA for apolipoprotein undergoes a deamination of a C, converting it to a U in such a way that a stop codon is created. This behavior is tissue-specific. The same kind of deamination, and the reverse U → C amination process, have also been observed in several plant mitochondrial mRNA transcripts. Moreover, even more unusual behaviors have been observed with mitochondrial RNAs. Bases can be deleted and inserted. The latter, especially, leads to a situation that can be interpreted as the formation of proteins for which there are no genes. In the most extreme case known to date, in the human parasite, *Trypanasoma brucei*, in the mRNA transcript leading to the formation of NADH dehydrogenase subunit 7 (a protein), as many as 551 Us are inserted throughout the transcript while 88 are deleted (Koslowski et al. 1990). In such a case it is hard to see why the DNA segment encoding such a transcript should be called the "gene for NADH dehydrogenase subunit 7."

(iv) Since the late 1960s it has been known that many segments of DNA are repeated many times in the genome. Moreover, many of these segments are known to be loci in the sense that they have definite coding or regulatory functions. Thus, the same classical locus may correspond to many different positions on different chromosomes.

(v) The genetic code which is supposed to capture the relation between gene and protein turns out not to be universal. Though the nonuniversality of the code is well-known, this still does not at present appear to be a particularly severe constraint by itself because the amount of known variation is not great.[60] At present the most extensive variations have been found in mitochondrial DNA in which, for instance, across all the major kingdoms UGA codes for tryptophan rather than terminate translation, as specified by the usual code. Mitochondrial DNA is special since mitochondria probably arose as independent organisms in biological prehistory and were subsequently incorporated into eukaryotic cells. Moreover, if prediction using the genetic code is all that is at stake, one could simply limit the use of the standard look-up table to nuclear DNA in eukaryotes. However, in at least four species of protozoa, UAA and UAG can code for glutamine rather than terminate translation even for nuclear DNA. Well across the various species that have been studied at this level so far, UGA is also known to code for amino acid residues that do not belong to the standard set of twenty. In the case of some viral DNA sequences, moreover, the UGA and UAG are sometimes but not always "read through," that is, ignored as termination signals. This happens within the same system, that is, in the same RNA sequence these codons sometimes result in termination and are sometimes read through. For example, the virus *Qβ*, which preys on *Escherichia coli*, has a coat protein that is usually produced by having UGA read as a termination codon. However, 2 percent of the time it is ignored, resulting in a longer coat protein whose presence turns out to be necessary for the normal behavior of the phage (Fox 1987).

6.6.   CLASSICAL AND MOLECULAR GENETICS

In spite of all the philosophical worries noted in § 6.4 and the complexities described in the previous section (§ 6.5), the relation between genetics and molecular biology are not hard to articulate. Consider Figure 6.6.1. It provides a scheme for the construction of the explanatory graphs introduced in Chapter 3, § 3.5.[61] Explanation, when successful, can only proceed from node to node along the edges, in the direction of the arrows. The hierarchy from **2** to **4** is the physical hierarchy of the organism, from the molecules upward. Node **4** is at the level of the individual organism; node **2** is at the level of the macro- and other molecules; in between are the cellular, tissue, and other levels. Strong (or physical) reductions occur along the (directed) edges of this hierarchy. The hierarchy from **1** to **3** is the genetical hierarchy. At node **1** are individual alleles. At node **3** is a (potentially complex) phenotypic trait. In between are loci, groups of loci, and simple traits. Continuous (directed) paths along this hierarchy describe genetic reductions of the type [type (iii)] that were discussed in Chapter 5.

The problems of the potential reduction of classical to molecular genetics that were discussed in the previous two sections concern only the tiny fragment of this graph. That fragment is the edge **b** and the nodes that it connects. The problems with the molecular definition of a (classical) gene discussed in the previous section suggest that explanations along **b** can be nontrivial. Nevertheless, they are possible, and when they become complex, they are fertile in the sense that they generate new directions for research. In particular, recalling the discussions in Chapter 2, § 2.5, note that definitions (if they are construed as being biconditionals in form) are not necessary for explanations. Thus, even though there may be no simple molecular definition of an eukaryotic (classical) gene, the rules operative at **1** can still be explained by those operative at **2**. As the discussion of the explanatory inadequacy objection in § 6.4 showed, and more detail will be given later in this section, there is no reason to suppose that this is not the case.

To reiterate and develop the point made there, classical genetics provided an account of the transmission of genes and a very rudimentary and rather unsatisfactory account of the expression of genes. The elucidation of molecular mechanisms of DNA replication, recombination, and cell division, especially meiosis (which is far from complete), begin the process of explaining the classical rules for gene transmission. Molecular accounts of gene expression are still rudimentary in eukaryotes. However, they not only explain but also correct classical claims, exactly as should be expected in a reductionist explanation. Two of the more important explanatory successes of molecular genetics in this context – the elucidation of mechanisms of DNA replication and recombination – will now be sketched.[62] These examples have been chosen both because they are critical to explaining the classical rules of gene

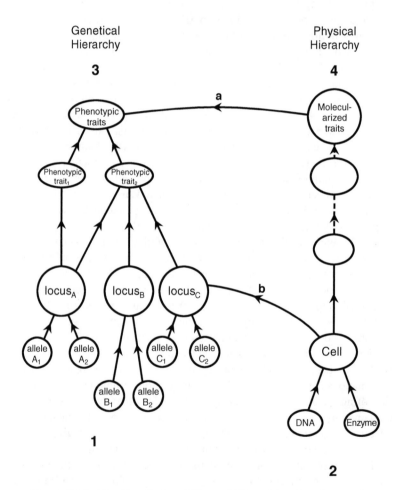

Figure 6.6.1. The Genetical and Physical Hierarchies. See the text for explanation.

transmission and because they are cases where simple cytological accounts do not provide the necessary explanation.[63]

The critical cytological interpretation of the physical processes that must underlie Mendelian inheritance are the processes of meiosis and fertilization. During meiosis, a single cell divides into four gametic cells, each with only one of the pairs of homologous chromosomes. Recombination takes place during this process. During fertilization two gametic cells, one from each parent (in cases of self-fertilization, both parents are the same), fuse to form a cell with the usual chromosome set. Assuming that the encounter and fusion of the germinal cells is a random event, and using the details of the meiotic process, both of Mendel's rules (as modified by linkage) can be explained. However, all of this, as already indicated, should be regarded (at least now) as cytological interpretation or redescription rather than explanation.[64] The molecular account of meiosis is necessary for a molecular reduction of the law of segregation of alleles; it also explains linkage and recombination. It, along with a molecular account of fertilization, is needed for the modified law of independent assortment (modified because of linkage). The additional assumption is that of random fusion – for this, no separate molecular account is necessary or, indeed, can be given.

Therefore, what needs to be explained – in order to get a reduction – is each step in these processes. Further, to be consistent with what has been said earlier in this chapter, it has to be shown that these explanations are those that use only the **F**-rules described in § 6.3. Consider meiosis first. The first stage is DNA replication.[65] The mechanisms of replication (in philosophical terms, which explain replication) are well-understood. These mechanisms are based on the structure of the DNA double helix, which consists of two antiparallel helices. One of these is said to be going in the $3' \rightarrow 5'$ direction while the other is said to be going in the $5' \rightarrow 3'$ direction; these directions are called the "polarity" of the helices. (The "3'" and "5'" refer to the orientations of the $3'$ and $5'$ atoms on sugar rings on the backbone of the helix.)

Five facts about the replication process are particularly interesting in this context: (i) it is initiated at several specific sequences, called "origins." Initiation often uses a short RNA fragment that is later excised and replaced by DNA. The Y-shaped structure is called a "replication fork." The separation of the DNA strands during this process, with each strand acting as a template for a new strand, explains why replication is semiconservative; (ii) fork movement is unidirectional or bidirectional; (iii) however, strands are always elongated in the $5' \rightarrow 3'$ direction; (iv) replication is generally semidiscontinuous. There is relatively continuous replication on the "leading" strand and relatively discontinuous replication on the "lagging" strand; (v) a variety of mechanisms control each of these processes, even within an individual cell. These mechanisms may depend on chromosome structure and conformation, the relative abundance of each enzyme, and so forth. They control the speed of replication, the continuity of the strands, the fidelity of the process, and so on. However, initiation takes place only once during a cell cycle. This helps explain why regular and predictable chromosome duplications take place during both meiosis and normal cell division (mitosis).

During meiosis, the stage after DNA replication involves the pairing of homologous chromosomes that line up on the spindle. As they begin to unpair, crossing-over of chromosomal parts possibly leading to recombination takes place. There are two models of recombination: the Holliday model and the Meselson-Radding model.[66] In the Holliday model (see Figure 6.6.2), after two homologous chromosomes are aligned next to each other, a strand of each chromosome with the same polarity is cut. The ends then leave the complementary strand of DNA with which they had been associated and join a complementary strand in the other chromosome. The resultant partially heteroduplex structure is called a "Holliday structure." In the Meselson-Radding model (see Figure 6.6.3) the same structure is generated by only one cut in single-stranded DNA, followed by DNA synthesis, the excision of a loop of DNA, the rotation of both chromosomes, and the migration of the branch point. In both models there is further migration of the branch point, leading to a continued transfer of a strand from each double helix (of one homologous chromosome of the pair) to the other. After rotation of the structure about the branch point (see Figure 6.6.4), depending on the details of enzymatic cleavage, two different end results can be formed. This already explains the recombination of alleles. More importantly, the two possible end results reveal a more complicated picture than what was envisioned in classical genetics. Moreover, in the heteroduplex structure there are likely mismatches between nucleotide base pairs in each double helix, making it unstable. Enzymatic correction of these can lead to one allele being lost and "converted" into another. Thus, the molecular models explain "gene conversion," a non-Mendelian process that had been observed especially in fungi but could not be subsumed under classical genetics.

In meiosis, after the first cell division, a second cell division takes place. The molecular mechanisms of these are fairly well understood. The critical observation, which is needed to explain the law of independent assortment, is that, in each set of four homologous chromosomes, which of them ends up in a particular daughter gametic cell is arbitrary. Similarly, during fertilization, which gametes fuse is shown by observation to be random. Random chromosomal assortment at the level of meiosis (after replication) followed by random fusion trivially explains the classical rules for the transmission of

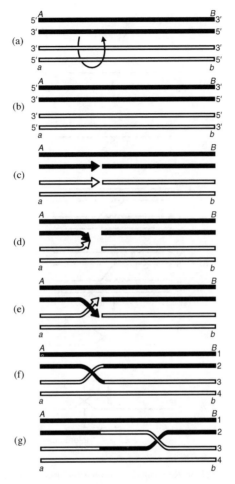

Figure 6.6.2. Holliday Recombination Model, Initial Stages [after Potter and Dressler (1978), p. 970; by permission of Cold Spring Harbor Laboratory Press]: (a) A pair of homologous chromosomes, each with a double helix, is shown. (Each helix is a bar of the diagram.); (b) The helices rotate along the arrow in (a) so that two adjacent helices from different chromosomes have the same polarity. (c) Two helices with the same polarity are cut. (d) The free ends at the cuts leave the other helix to which they were each hydrogen-bonded. (e) Each of these free ends now becomes associated with the corresponding helix in the other chromosome. (f) Ligation of the free ends results in a partially heteroduplex structure (the Holliday structure). (g) The branch point migrates, leading to a continued exchange of DNA strands. Later stages are shown in Figure 6.6.4.

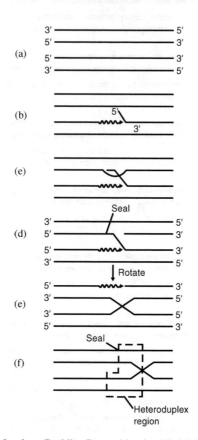

Figure 6.6.3. Meselson-Radding Recombination Model, Initial Stages [after Darnell, Lodish, and Baltimore (1990); by permission of W. H. Freeman]. A pair of homologous chromosomes, each with a double helix, is shown. (Each helix is a line of the diagram.) (a) Adjacent helices from different chromosomes have the same polarity. (b) One helix is cut, a strand displaced by DNA synthesis at the 3′ end. (c) The displaced strand is taken up by the adjacent helix on the other chromosome, leaving an unpaired loop. (d) The loop is excised, leading to a single-stranded "invasion." (e) Rotation of both chromosomes produces two cross-over strands. (f) Branch migration creates a heteroduplex structure (the Holliday structure). Later stages are shown in Figure 6.6.4.

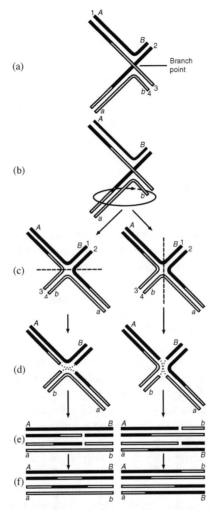

Figure 6.6.4. Recombination Model, Later Stages [after Potter and Dressler (1978), p. 970; by permission of Cold Spring Harbor Laboratory Press]. The later stages are the same for both the Holliday and the Meselson-Radding models. (a) Branch migration continues. (b) Rotation takes place in the direction of the arrows. (c) In the rotated structure, enzymatic cleavage can occur in the two ways shown by the broken line. (d) The two possible results of enzymatic cleavage; the subsequent patterns of rejoining are shown by the dotted line. (e) The two possible results of the rejoining. (f) The two possible end-structures after rejoining. Note that classical accounts of crossing-over did not distinguish between these two possibilities. Molecular genetics thus adds detail to the classical picture.

genes. Thus, the statistical independence assumptions that are incorporated in Mendel's laws are not removed in molecular genetics. What has happened is that they have been transferred from the abstract level of genes to the level of tangible observations of the dynamics of cellular or subcellular entities. The process of fusion during fertilization, which will also not be discussed here, is also understood in detail and varies slightly between plants and animals. Nevertheless, there are no puzzles here. What molecular explanation has provided are physical accounts of all these processes. Moreover, the two mechanisms discussed in detail here, DNA replication and recombination, both explain phenomena that were not (explanatorily) accessible even at the cytological level.

The following four points about the preceding discussion of mechanisms should be emphasized.

(i) Even the best-characterized mechanisms, such as those of replication or recombination, at the present level of detail, still do not give complete explanations of the processes. For instance, more complete answers are required to the questions of what initiates replication or recombination. However, this cannot be regarded as a sign of explanatory inadequacy if classical genetics is what is to be explained: classical genetics had nothing at all to say about the reasons for the initiation of these processes.

(ii) More importantly, this type of explanatory incompleteness provides molecular genetics with a problem domain for further exploration. It is not a result of the presence of anomalies. Instead, all that it shows is that there is more to be deciphered and solved to have complete molecular explanations even of replication and recombination (let alone mutation and gene action). It keeps these reductions fertile, one of Nagel's (1961) most important criteria for the scientific significance (and philosophical interest) of reductions.

(iii) Because of the complexity of the interactions at the molecular level, the molecular explanations provide richer accounts of the phenomena than what classical genetics did. For instance, there are varieties of recombination. There is an account of the process by which recombination takes place, showing why the probability of recombination is not simply a linear function of distance between loci on chromosomes. Though it will not be discussed here, molecular genetics similarly provides a rich account of mutagenesis, about which classical genetics had little to say.[67] Thus, these reductions extend, correct, and reveal a richer understanding of reproductive phenomena than what was possible without the reductionist enterprise.

(iv) Nothing has so far been explicitly said about the role of the **F**-rules from § 6.4. However, they have been implicitly assumed – and they are all that has been assumed – in the various enzymatic and other interactions mediating the mechanisms that have been described. At some stages of these processes, precisely how these rules regulate the mechanisms have not been fully elucidated; that is the important reason why there still is explanatory incompleteness. In fact, establishing the sufficiency of these rules would complete the explanatory project.

Returning to Figure 6.6.1, it is important to note that the fragment, **b**, is only a minor part of the total explanatory scheme. Usually, what is desired are explanations of phenotypic behavior at node **3**. If these explanations are to originate at the molecular level, that is, at node **2**, there are two pathways that can be followed. One is the $2 \to 4 \to 3$ pathway. Classical genetics is entirely circumvented in this pathway. If the rules operating at **2** are taken to be those of molecular genetics, then this strategy would be one in which molecular genetics replaces rather than reduce classical genetics, as suggested by Hull (1972, 1974). Problems about reductions can occur along this pathway. There are two types of such problems.

(i) The molecular characterization of phenotypic traits, by definition or by explanation, as represented by the edge **a** can be problematic. For instance, in humans, while the sickle-cell trait or PKU might have a simple molecular characterization (at the phenotypic level), the problem of characterizing homosexuality or schizophrenia at the molecular level is at present entirely unsolved. In fact, in such cases the physical characterization of the trait at any level (of the physical hierarchy) is far from simple. It may even turn out that there may be many different physical states of the nervous system that correspond to what is called schizophrenia, and that the molecular mechanisms leading to them, as Hull seems to suggest, may be varied. This leads to the second problem with carrying out reductions along the $2 \to 4 \to 3$ pathway.

(ii) There can be problems in carrying out the explanations from **2** to **4**. However, this problem is no different than the usual problems about reduction that occur in any physical context. There is nothing peculiarly biological about it. In fact, recalling the examples treated in § 6.4, what is surprising is how much reduction can be carried out. From a physical point of view, so far biological systems have been conceptually rather simple: they have, indeed, lived up to the machine metaphor to a much greater extent than many of the systems encountered in physics.

The other is the $2 \to 1 \to 3$ pathway. Once again, there are two types of problems that can occur along this pathway.

(i) The problems along **b** that have already been discussed above may occur. While important, they do not undermine the explanatory project of molecular genetics.

(ii) The problems on the way from **1** to **3** may also occur. The latter is the problem of explaining phenotypic traits from a genetic basis – the problem of genetic reductionism that occupied Chapter 5. What is important to note is that the latter problem and any of the problems associated with the $2 \to 4 \to 3$ pathway are irrelevant to the reduction of classical to molecular genetics. This point has far too often been ignored in the antireductionist literature. Hull's (1972, 1974) account of multiple realizations of classical processes deals almost exclusively with problems along the **2** to **4** pathway. Most of Rosenberg's (1985) worries are about the pathway from **1** to **3**.[68] So the relevance of these objections to that reduction is less than obvious.

6.7.   THE DECLINE OF DOMINANCE

The picture of (strong) reduction in molecular biology that was outlined above (in § 6.2 and § 6.3) may appear to be so liberal, particularly to those who are accustomed to viewing reduction according to the formal schemes described in Chapter 2, that reduction might appear to be fully ensured in molecular biology since it begins with a physical representation of a biological system. How, conceivably, could strong reduction of this sort fail? It can be the case that a system can never be given a hierarchical representation for explanatory purposes. This is a standard situation in quantum mechanics where properties cannot be attributed to one constituent of an interacting system without referring to all others.[69] Around the turn of the century, as noted in § 6.1, physiologists made the same claim for biological systems. Those claims turned out to be misguided, and quantum mechanics is explanatorily irrelevant to molecular biology (§ 6.3). Nevertheless, there is ample empirical reason to be open to the possibility that reductionism may eventually fail in molecular biology. There is one trivial and relatively uninteresting sense in which it fails: molecular biology involves modes of explanations, such as functional explanation, that are not (or at least not straightforwardly) reductionist.[70] But there is also a stronger sense in which it may fail, and this is much more interesting.

Attempts at reductionist explanation may fail even when the mode of explanation is straightforwardly "efficient," that is, it relies directly and entirely on preexisting factors (unlike the case of functional explanations, for instance, which rely on future functions). How could this happen? Two examples will illustrate the potential for such a failure. In the process, they will also serve to emphasize, again, the point that whether reductions are going to be successful or not – broadly speaking, reductionism – is an empirical issue.

(i) The first example is one of the best-known unsolved problems in molecular biology, the "protein-folding problem," that of predicting the three-dimensional conformation of a protein molecule.[71] As early as 1958, Crick had put forward the sequence hypothesis, that the linear order of amino acid residues in a protein (later called its "primary structure") would determine its conformation (later called its "tertiary structure"). Forty years later, in spite of the astronomical increase in computational ability, the sequence hypothesis remains unproven, not only for all proteins, but even for any single one. Even the conformation of as simple a protein as bovine pancreatic trypsin inhibitor (BPTI) with only fifty-eight amino acid residues cannot be predicted or explained from its sequence and the **F**-rules of macromolecular physics. Most "typical" proteins have between 100 and 500 residues; many have several thousand.[72] It is usually believed that the problem is with the sheer complexity of the computations involved, whether the preferred conformation is sought by minimizing the energy of the protein, or by tracing the dynamics of the individual atoms. However, it is also clear that (i) specific "pathways" or (ordered) sets of conformations may be involved in protein folding;[73] and (ii) the folding process is critically mediated by specific molecules called "chaperones," which also act to enable proteins to retain their "functional" or "normal" conformations within cells.[74]

There are at least two different ways in which the sequence hypothesis could be false: (a) it could be the case that more than the linear order of the amino acid residues is required to explain protein conformation. There is already considerable evidence for this. For instance, the history of protein formation at the ribosomes could be relevant. This would be one way in which a folding pathway is generated. (The evidence against this possibility is that many proteins, after being denatured or "unfolded" (for instance, by heat) still regain their original conformation.) Moreover, the apparently ubiquitous role played by molecular chaperones is a strong argument against the sequence hypothesis. Clearly, if chaperones are necessary, then its sequence alone cannot explain the conformation of a protein.[75] Or, (b) the **F**-rules of macromolecular physics may not be adequate. Should the sequence hypothesis turn out to be false because of the second of these alternatives, physical reductionism will have failed in this case. However, a failure of the sequence hypothesis because of the first alternative is not necessarily a failure of physical reductionism. There may be more to protein folding than sequence information alone, and these might all be mundane macromolecular physical processes.[76]

(ii) A more troublesome and interesting case is that of dominance. The phenomenon of dominance, that for some traits the heterozygote is phenotypically indistinguishable from one of the homozygotes, was recognized by Mendel. All seven traits that he studied exhibited dominance. When Mendelism was rediscovered around 1900, it became clear that dominance was not exhibited by all traits.[77] Some traits showed no dominance, that is, the heterozygote was exactly intermediate to the two homozygotes. Others exhibited various degrees of dominance, including "overdominance" (when the heterozygote exhibits a trait to a greater extent than either homozygote).[78] Dominance became an important part of the conceptual framework of classical genetics. If a trait affected fitness in a continuous manner, the degree of dominance could be correlated with fitness.

In the 1920s and 1930s the origin of dominance became a matter of controversy among evolutionary biologists. Fisher (1928a,b) argued that it was an evolved property that had emerged because of selection.[79] Wright (1929a,b) doubted that there was such a selectionist story to be told. From his point of view, dominance was a result of ordinary physiological or biochemical interactions between genes and their products. Haldane, Muller, and others positioned themselves somewhere in the middle. When described in these general terms, Fisher's and Wright's positions appear to be potentially complementary, rather than in conflict: it is possible that dominance could have been selected for, and arises in each individual through common biochemical mechanisms. However, the specific model of selection that Fisher assumed, in which several loci were involved, was not compatible with Wright's: the mechanisms that were compatible with the two positions were different.

It would appear that molecular biology, with its emphasis on the elucidation of mechanisms of biological interaction, would help decide the dispute between Fisher and Wright. What happened instead is that, in molecular biology, perhaps because of its origins in the study of haploid organisms, the concept of dominance disappeared. No biologist, and only a few philosophers, have supposed that molecular biology provides a simple account of dominance. Interpreting dominance as a property of an allele, rather than one of a trait (as is commonly done), Schaffner (1969) attempted to define "dominance" of a gene (allele) by its capacity to direct the synthesis of an active enzyme. Hull (1974) correctly pointed out that this was implausible simply because dominance is a relational property (an allele must be dominant with respect to another). He also noted the complications about the various types of dominance that were noted before. Schaffner (1993a, p. 442) concedes Hull's first point but continues to defend the definition:

Allele a is dominant (with respect to b) =
(DNA sequence α, DNA sequence $\alpha \Rightarrow$ amino acid sequence **N**) &
(DNA sequence β, DNA sequence $\beta \Rightarrow$ amino acid sequence **M**) &
(DNA sequence α, DNA sequence $\beta \Rightarrow$ amino acid sequence **N**).

The "$\Rightarrow$" in the definition is supposed to be spelled out using molecular mechanisms. There are at least three reasons why this attempted definition is irrelevant.

(a) It simply assumes that dominance is to be judged at the level of the amino acid sequence. In other words, it endorses the claim that phenotypic traits and differences must be characterized at that level (at least in discussions of dominance). While molecular biologists have also sometimes endorsed such a position,[80] it makes a parody of genetics. Genetic rules and explanations are powerful because there are many traits far removed from DNA that can be explained and understood from the presence of particular alleles at a single locus (or at a very few loci). If explanations from the rules of genetics cannot be pushed any further than the level of the amino acid sequences, genetics would hardly be as powerful and as significant as it is usually supposed to be. When a molecular explanation of dominance emerges – if it does – it must account for dominance at the level of polydactyly or Huntington's disease. That is the interesting challenge for molecular biology.

(b) There are many cases known when both alleles are expressed at the amino acid residue sequence level but the trait (or the relevant allele) shows dominance. For instance, in humans, sickle cell disease is recessive even though both ordinary (HbA) and sickle cell hemoglobin (HbS) are present in about equal amounts in the heterozygote (see § 6.2), showing that both alleles are equally expressed at the amino acid residue sequence level. Thus, there is dominance at the gross phenotypic level but not at the molecular level to which Schaffner confines dominance.

(c) Perhaps even more importantly to the extent that dominance is understood at the molecular level, that understanding is in complete variance with Schaffner's scheme.[81] The basic – and controversial – model was outlined in a paper by Kacser and Burns (1981). This model assumes that what is to be explained is the recessivity of new mutations, which already imposes a restriction on its potential domain of applicability.[82] Kacser and Burns characterize the organism as a biochemical system, more specifically, "as an enzyme system [which] consists of a large array of specific and saturable catalysts organized into diverging and converging pathways, cycles and spirals all transforming molecular species and resulting in a flow of metabolites" (p. 641). Though the discussion is couched in terms of enzymes, any network of catalytic molecules satisfies the assumptions of the Kacser-Burns model. The net flux across such a network is "closely related to the characters or phenotype" (p. 641). If J is the flux, and e is the concentration of an enzyme, E, the flux control coefficient, C_e^J, defined by

$$C_e^J = \frac{\partial \ln J}{\partial \ln e} \qquad (6.7.1)$$

measures the effect of E on the flux. The crucial result, first proved by Kacser and Burns (1973) is that, for many systems of this sort, there is a "summation theorem":

$$\sum_e C_e^J = 1. \qquad (6.7.2)$$

Suppose that all the C_e^J are positive, and that the number of different enzymes in the network is large. Then the C_e^J are each small. In such a situation, even a large change (say 50%) in the activity of an enzyme at a particular junction of the network due, for instance, to a mutation would still have a very little effect on the net flux (J) and, therefore, on the phenotype. This is the "molecular basis" of dominance. It does not require, contrary to Shaffner, that only one of the alleles at a locus gets expressed as an amino acid residue sequence.

The basic Kacser-Burns model assumes a linear chain of reactions. In that case the assumptions about the C_e^J used in the argument of the previous paragraph are justified, and dominance is an inevitable property of such reaction networks.[83] What if the reaction networks are branched? Kacser and Burns (1973) reported simulations that led to the same result but these simulations only explored a few architectures, mainly

172

those in which the branches eventually converged on the original pathway (forming cycles). Branching is important because it can produce negative values for the C_e^J when the (local) flux from a branch is directed inwards. When such C_e^J are large, or when there are many branches with negative C_e^J, the value of the C_e^J for a particular enzyme may make a large contribution to the net flux even if the summation theorem [Equation (6.7.2)] holds.[84] In such circumstances the Kacser-Burns argument does not go through and, in fact, there may be no dominance: the effect of changing one enzyme may have a significant effect on the net flux and, therefore, the phenotype.

Thus, whether or not dominance arises is only minimally explained from the properties of individual enzymes. The explanatory weight falls on the *topology* (or architecture) of the network, that is, its connectivity properties, especially how much branching there is, whether the branches eventually return to the main pathway, and the direction of the (local) fluxes along these branches. For want of a better name, this pattern of explanation will be called *topological*. These explanations are not physical reductions. The structural assumptions which, in this context, bear the explanatory weight are not the **F**-rules from macromolecular physics. Rather, they are assumptions about topology that are naturally represented as weighted directed graphs: each enzyme becomes a node, all DNA segments are sources (in general), and the end-products are sinks. The C_e^J provide the weights of the edges; their signs provide directions. Calculating the flux becomes a flow problem from sources to sinks. It is not being suggested here that the Kacser-Burns model provides the correct explanation for dominance. But it might be. If it is, it constitutes a clear case where ordinary (nonfunctional) explanation in molecular biology would still not be reductionist. It shows that (physical) reductionism, as construed in this book, is not a trivial or empty thesis. It may be false – whether, and to what extent, it is true is an empirical question.

The nature of topological explanation is beyond the scope of this book. Suffice it simply to note that there are many other situations in which topology might absorb the explanatory weight of explanations. Restricting attention to molecular biology, perhaps the most interesting of these is the immune (or idiotypic) network theory, where the regulation of the immune response is explained from the connectivity properties of networks of interacting antibodies (on B-lymphocytes).[85] Outside molecular biology, topology rears its intriguing head in many situations including connectionist models of the mind, models for the evolution of genetic networks,[86] or even development.[87] Topological explanation deserves more attention than philosophers have so far accorded to it.

6.8. CONFLATIONS

It is important not to conflate a genetic reduction, as described in Chapter 5, with a physical reduction, as described here. Perhaps even more importantly, genetic reductionism should not be conflated with physical reductionism. From the point of view of physical reductionism, DNA enters the molecular milieu on par with proteins or, for that matter, lipids or any other molecules that are found in living organisms. Physical reductionism does not require any assumption about the primacy of DNA or of genes in the explanation of biological behavior. It does not require the non-inheritance of structures other than the DNA sequence. Ultimately, physical reductionism is a thesis about the relation of biology to physics; it is neutral about the status of genetics within biology. (It is also a very interesting thesis about physics – about the physics of systems organized very differently from those conventionally studied in statistical physics.)

Genetic reductionism was described in Chapter 5. This is the thesis that genes can explain all phenotypic features of an organism, from the origins of its physical characteristics to the details of its behavior, including complex social behavior and, for humans, even psychological and mental behavior. It is a thesis of remarkable scope. However, its general validity is doubtful (at best). Genetic reductionism naturally leads to an emphasis on DNA, which specifies the alleles, over all other biological molecules. In recent years it has led to a deification of the DNA sequence, the most public manifestation of which is the Human Genome Project (HGP), the attempt to sequence what is supposed to be a "consensus" or "representative" human DNA sequence.[88] The proponents of the HGP have promised many advances in biology and medicine. They have largely relied on the success of molecular biology – the power of molecular explanation – to justify their faith and to convince political agencies responsible for the disbursement of public funds. However, as this chapter has demonstrated, these successes are primarily successes of physical reductionism. Nevertheless, in a rather remarkable example of conceptual conflation, the proponents of the HGP have used these successes to justify a general thesis of genetic reductionism.[89] It is unlikely that the HGP will deliver on almost any of its promises, scientific or medical, but the problems there are not with physical reductionism. Rather, they are the problems that have plagued genetic reductionism since the 1920s and were discussed in the previous chapter.

7

Concluding polemics

This book has attempted to explicate three different types of reduction (and reductionism) that have been relevant to genetics and to assess their relative strengths and weaknesses. The conclusions reached for each of these types have been stated in each of the previous three chapters. No further conclusion is strictly necessary, though some further consequences of those conclusions will be listed in the third (and final) section of this chapter. Meanwhile, the first two sections of this chapter will explore some topics that deserve more attention than what has been given to them so far. In particular, recalling that the stage for the discussion of reductionism in genetics was set at the beginning of this book by noting how claims that various traits are "genetic" have become pervasive in recent years, an attempt will be made here to define that term.

To develop the background necessary for that proposed definition, § 7.1 will draw some distinctions about biology, genetics, and inheritance. These are relatively trivial but, as an example given there will show, they have far too often been ignored during the recent excitement over the prospects for genetic reductionism. This will set the stage for § 7.2 where a (rather restrictive) proposal for the use of the term "genetic" will be made. Such a proposal is important mainly because of a social context in which claims of the genetic basis for traits are being made with increasing frequency (see Chapter 1) and, especially in the popular press, with little concern for any of the subtleties or nuances that were emphasized in Chapter 5, § 5.5 and § 5.7. The success of this attempt is not necessary for the analyses of the previous three chapters to have been useful. However, this attempt is partly based on those analyses though it does not follow from them in any strict sense. Finally, § 7.3, which comes closest to a conventional conclusion, notes some of the more salient points that do mainly follow from those analyses.

On occasion in this chapter, the tone will be polemical with respect to philosophical issues in general and to some interpretations of the state of the current knowledge of genetics. It should be emphasized that these polemics are not entailed by the analyses of the preceding chapters. Moreover, an effort has been made to ensure that the scientific discussions are kept as free of these polemics as possible.

7.1. GENE, ORGANISM, AND ENVIRONMENT

Not everything biological is genetic, though the emphasis on genetics that has permeated twentieth-century biology almost suggests the opposite.[1] Its most recent manifestation is the Human Genome Project (HGP) that, as even its most ardent advocates concede, has shifted scarce resources away from other biological research, at least in the United States.[2] The next section of this chapter will attempt to restrict the use of "genetic" only to traits that are most appropriate. To set the stage for that discussion, this section will lay out some distinctions between "biological," "genetic," and "inherited." These will also be distinguished from "heritable" in the sense of the technical notion of heritability though, after what has been said in Chapter 4, this is hardly necessary. The distinctions that will be made here are generally trivial. Nevertheless, as an example discussed at the end of this section will show, they are sometimes missed in unexpected quarters.

To distinguish "genetic" from the other categories, it is useful to recall how alleles act in an organism. There are two major ways: (i) an allele may code for a polypeptide chain, that is, a protein or a part of a protein. This is called the expression of an allele. An intermediate stage in this process is the production of RNA from that allele. In the rest of this chapter the polypeptide and RNA chains coded for by an allele, if any, will both be called "immediate products" of that allele. (ii) through a variety of mechanisms, an allele can regulate the expression of other alleles. It can, for instance, be at a regulator locus itself (as in the case of alleles of the regulator locus (o) in the *lac* operon discussed in Chapters 3, § 3.6 and in Chapter 6, § 6.2). Alternatively, the RNA chain that it codes for can play a role in regulation. The important point to note is that an allele can play these roles without being expressed as a polypeptide chain.

Whether alleles act in any way at all depends on the chemical environment of the chromosome on which that allele is located; that is, chemical cues from the environment regulate the action of any allele. In the very first stages of development, these chemical cues often originate from the mother, in the cytoplasm of the fertilized egg. Some of what is inherited in the cytoplasm might be genes (in mitochondria or chloroplasts, for instance) but most of the chemicals that act as cues for the genome of the cell are environmental in the sense that they are not alleles or immediate allelic products of the incipient organism's own genome. All of what is inherited through the cytoplasm contributed by the mother is called "cytoplasmic inheritance." While the inherited cytoplasmic chemical cues play important regulative roles on the alleles during the earliest stages of an organism's development, the cellular environment of the genome eventually comes to depend increasingly on the organism's external environment that provides it with sustenance. The pathways by which the external environment impinges upon the cellular environment of a set of alleles that can potentially be expressed can, of course, be exceedingly complex. However, that complexity does not in any way decrease the role of the external environment in the expression of an allele.

For the normal development of an individual organism, the time of action of the alleles is critical. The complex regulatory processes that ensure this are based on the interaction between the environment and the alleles (that is, the genotype of the organism). This is why the alleles alone are not sufficient for the occurrence of any trait. Because of the complexity of the environment–genotype interactions, there can be tremendous variation in the degree to which any allele is expressed. As already discussed in Chapter 5 (§ 5.3), this degree is the expressivity of the corresponding trait; similarly, the (conditional) probability of the trait being expressed at all, given the genotype, is its penetrance.

With even this little biology, it is already possible to make some important distinctions, those between "biological," "genetic," "inherited," and the technical concept of "heritable." These distinctions should be emphasized by themselves for two reasons. First, they should be maintained even if the particular definition of "genetic" that will be proposed in the next section turns out to be flawed. They are independent of that definition, and they already show that one should not infer (or assume) that some trait is genetic simply because it appears to be biological or even because it is inherited. Second, the ability of the proposed or any other definition of "genetic" to make precise and clarify these preliminary distinctions can serve as an important desideratum for judging the adequacy of that definition.

The distinction between "biological" and "genetic" is clear enough. Even without trying to define "biological," it should be uncontroversial that an organism is not purely the result of its genotype, or of its environment, but of the history of the interactions between its genotype and its environment. Individual organisms differ partly because of differing genotypes, partly because of the different environments in which they develop. The genotype and the environment interact to produce the organism. Moreover, the genotype–environment interaction is historically contingent in the sense that what matters is the precise temporal sequence of the events that comprise that interaction. For instance, in humans, exposure to the amino acid residue phenylalanine in the diet leads to mental retardation in individuals homozygous for the PKU allele if it occurs during the first six years of life. In later life a diet containing phenylalanine becomes at least partly acceptable.[3] Indeed, it is precisely because of the gene–environment interaction that attempts to define "genetic" by invoking the sufficiency of a set of genes faltered, as noted earlier (in Chapter 1, § 1.1).

Consequently, a characteristic of an organism can be biological without being genetic. Mental retardation due to PKU is obviously biological. It turns out that it is also genetic (informally, and according to the definition that will be proposed in the next section). However, that is an independent point. Physical or mental retardation (in humans) due to malnutrition is, similarly, obviously biological. It is also, at least superficially (though a precise defense of this claim will have to await the definition of "genetic" to be given below), not genetic since it can occur to any individual irrespective of genotype. Turning to a potentially much more complicated example, even if it is demonstrated that there is some (biological) mechanism by which different morphological properties of brains lead to different sexual behaviors or orientations (for instance, in humans), all that it would establish is that these sexual behaviors are biological.[4] It would not, by itself, establish that they are genetic: the morphological development of the brain could well be due to a myriad of environmental factors just as certain physical characteristics of ambitious athletes can be enhanced by steroids, or both the physical and mental development of children can be retarded through malnutrition.

The distinction between "genetic" and "inherited" is equally important. As noted above, an organism inherits from its parents more than just its genome. Moreover, in "higher" organisms, capable of manipulating their environments, parents might ensure uniformity of environment from generation to generation. There are many cases of cytoplasmic inheritance where some property of an individual is not under the control of its own genotype. The classic example is the chirality of spiral coiling of the snail, *Limnaea peregra*.[5] Most individuals of this species are dextral, that is, their shells and bodies are coiled in a right-handed direction. Some individuals, however, are sinistral, that is, their coils are left-handed. It turns out that what controls the chirality of an individual is not its own genotype but a single locus of its mother. Sinistrality is recessive: if the mother is homozygous for the relevant allele, the offspring shows sinistral coiling irrespective of whether it is itself homozygous for that allele. What happens is that some cytoplasmic molecules produced by the expression of the mother's alleles persist in the fertilized egg and confers sinistral chirality to the offspring. If the maternal grandmother was also homozygous for the allele for sinistrality, then the mother would also have had sinistral coils. In such a circumstance sinistrality would be inherited in the sense of being passed from mother to offspring without the offspring necessarily inheriting the mother's genotype at that locus. All that this shows is that the relationship between "genetic" and "inherited" is not straightforward: a trait can be inherited without the allele responsible for it.

Conversely, it is also possible that a certain property or trait is genetic in the sense of being under the direct control of a few loci but not inherited. For example, mutations often alter the genotype of specific cells of plants. These are only inherited if these cells end up as ancestors of germinal cells for the next generation. In many "higher" animals, germ-line sequestration ensures that somatic mutations or other changes in the genotype of cells other than those in the germ-line are not inherited. Indeed, these mutations and a high degree of somatic cell recombination are the mechanisms by which the mammalian immune system maintains the remarkable antibody diversity that it uses to respond to any external agent. These mutations and other genomic changes are not inherited (by the individual organisms in the next generation) but the traits that they give rise to, the specific antibodies, are certainly genetic in the sense that there are precise alleles at specific loci for them.

Finally, there is an important distinction between what is "inherited" and what is "heritable" in the sense of having a high "heritability" according to the technical definitions of heritability discussed in Chapter 4. At the expense of reiterating a point that was made many times there, the value for either narrow or broad heritability for a trait might say nothing about its inheritance. A simple example will suffice here: consider the human trait of having two legs. The variance of the genetic component of this trait is 0 (or at least almost 0) since the human genome is such that all individuals seem to carry whatever alleles that are necessary to give two legs to humans. The total variance, however, is nonzero since some individuals lose one or two legs during their lifetimes, such as those who are victims of wars. This gives a narrow heritability and, therefore, also a broad heritability of about 0. It should be obvious that this technical concept of heritability says little about what is inherited. Conversely, in a very constant environment and a genetically heterogeneous population, any environmentally controlled trait will have high heritability.

These distinctions are simple enough, perhaps so simple that it might seem that they are being belabored for no real purpose. However, they are lost often enough, even in relatively sophisticated scientific circles. Consider as an example, an editorial from *Science* from within the last decade.

Last week a crazed gunman terrorized hostages in a bar in Berkeley, killing one and wounding many others. Homicidal maniacs have appeared in all cultures over the entire length of human history. Society's modern response to their chaotic behavior has most often been a diligent search of their childhoods, as though understanding their upbringing and circumstances would explain their aberrant actions. There is nothing wrong with that kind of investigation, and in some cases history and environment will reveal clues. However, it is time the world recognized that the brain is an organ like other organs – the kidney, the lung, the heart – and that it can go wrong not only as a result of abuse, but also because of hereditary defects utterly unrelated to environmental influences. ... [T]he Human Genome Project ... will provide information of particular value to the study of the neurosciences. ... There are legitimate arguments in regard to how fast such a project should go or how it should be administered, but there seems to be little doubt that it will help in the mental health area. Schizophrenia (the disease from which the Berkeley gunman is thought to have suffered) and other mental illnesses can have a multigenic origin. A sequenced human genome will be a very important tool for understanding this precise category of diseases" [Koshland (1990), p. 189].

Leaving aside the controversial question of the sense in which complex mental behavior can properly be termed "biological," what is remarkable in this passage is the ease with which a transition is first made from the biological to the inherited (if there is a biological basis for "insanity," it is likely to be "hereditary") and, then, from the "inherited" to the "genetic" (hence the relevance of the Human Genome Project). Of the distinctions elaborated in this section, that between "inherited" and "genetic" is explicitly, and that between "biological" and "genetic" implicitly ignored. So far, only the technical notion of heritability seems to have escaped conflation. However, as if to make amends for this, the editor continues (after lyricizing the claims of the proponents of the Human Genome Project).

As we extend the life expectancy of individuals and provide cures for infectious diseases, the affliction of mental disease becomes more glaring. Advancing research can cure some fraction of these illnesses. It may also provide predictive diagnoses to distinguish those who are severely ill from those who merely represent harmless aberrations from the norms of society. The article on identical twins reared apart shows that some physiological and psychological traits are inherited; however, this does not minimize the influence of environment and motivation. While some inherited illnesses cannot be alleviated without a biochemical cure, in others there is only a tendency to disease, which can be ameliorated or prevented by a helpful environment" [Koshland (1990), p. 189].

While there is here a welcome appreciation of some of the complexities of the heredity–environment interaction, nevertheless, there is an explicit conflation of what is inherited with the technical notion of heritability and an implicit one between what is "heritable" and the "genetic." The study of identical twins, which is alleged to show what is inherited and genetic, was the work of Bouchard et al. (1990), which consisted entirely of an analysis of broad heritability.

7.2.   WHAT'S "GENETIC"?

With these distinctions in mind, it is now possible to propose a definition of "genetic." Before such a definition is offered, however, it is useful to identify some minimal criteria that any such definition should satisfy. These should be regarded as adequacy conditions for any attempted definition of "genetic." There are three such criteria, two of which have already been mentioned: (i) the definition should be such that the traits that have routinely and uncontroversially been designated genetic remain so; (ii) an adequate definition of "genetic" should be consistent with the distinctions discussed in the last section; and (iii) the definition should at least illuminate, if not fully underscore, the difference between the respective roles of genes and environment that have been routinely disputed as part of the apparently perennial nature–nurture controversy (see Chapter 4, § 4.4). These three criteria are of decreasing importance. If (i) is not satisfied, there is hardly any reason to call the proposed definition a definition of "genetic"; if (ii) is not satisfied, "genetic" would lose much of the ways it is naturally used in ordinary biology. However, as will be emphasized below, (iii) may prove to be impossible to satisfy in many cases simply because nature and nurture may be inextricably entangled in the phenogenesis of certain traits.[6]

The strategy followed here will be to call a trait "genetic" if and only if its occurrence is best explained on the basis only of the properties of genes (specific alleles at specific loci). Thus, this strategy is parasitic upon the account of genetic reductionism given in Chapter 5, and also partly upon that of physical reduction in molecular genetics (pursued in Chapter 6). Note that this strategy permits traits not to be genetic; genes may not be the best explanation for a trait, or may not even explain it at all. It also avoids all ontological issues of genetic determinism, and so on, and does not deny the relevance of environmental factors even for genetic traits. The proposed definition will be more specific than simply requiring that the "trait be reduced to the genes." It will draw on the idea that this can only be done provided the number of loci involved is very small and their roles in phenogenesis relatively unproblematic. Thus, it will draw on the discussions of Chapter 5, though it is not entailed by that discussion.

The proposed definition is that a trait be called "genetic" if and only if the following three conditions are satisfied:

(i) the trait is under the control of a few loci;

(ii) the trait always (that is, in all populations) shows high expressivity; and

(iii) the immediate products of the alleles at these loci form part of the biochemical characterization (that is, description at the biochemical level) of the trait.

If conditions (i) and (ii) are satisfied, a trait will be said to be "under genetic control." It might also be genetic, but that will require the additional satisfaction of condition (iii). The use of "under genetic control" to refer to the status of certain traits was standard in classical (that is, premolecular) genetics. Since condition (iii) could not be considered in a situation when even the chemical nature of the gene was unknown, classical genetics generally relied on conditions (i) and (ii). Thus, this terminology is consistent with what was standard in biological practice in the premolecular era.

Note that there is no explicit requirement of high penetrance in this definition. The penetrance of a trait (given a genotype) depends on the population (and, therefore, on the environment) simply because of its definition as a conditional probability (of the trait being exhibited, given the genotype) defined for a specific population (see Chapter 5, § 5.5). An attempt is being made here to prevent the definition of "genetic" from becoming an irreducible property of a population (in the sense of Chapter 4). If a requirement of high penetrance were explicitly incorporated into the definition of "genetic," that definition would become irreducibly dependent on the population. Therefore, there is no explicit requirement of high penetrance. However, implicitly, a high penetrance is assumed because, if a trait controlled at a single locus (or a small number of loci) is always to have high expressivity, its penetrance must be close to 1. Similarly, no assumption is being made about the presence of phenocopies. As explained in Chapter 5 (§ 5.6), their presence would only permit a trait localized to a given pedigree (or an appropriately larger subpopulation) to be genetic (or under genetic control), without the trait in general being genetic (or under genetic control).

Two aspects of this proposed definition should be immediately apparent.

(i) It is highly restrictive. Whether it is too restrictive will be discussed later in this section when the question of the satisfaction of the first adequacy condition is taken up.

(ii) There are two points of ambiguity in the definition: (a) how many are a "few" loci?; (b) how high a degree of expressivity is required? Ultimately, for a satisfactory definition, these ambiguities will have to be removed or at least decreased to a tolerable level. However, there is no point in expending efforts toward that end until it is clear that the proposed definition is at least acceptable as far as other issues are concerned. The problem of these ambiguities will, therefore, be set aside. If this proposed definition is accepted, future work will have to determine what values for these parameters should be regarded as adequate. That will require the adoption of even more conventions than those already required by the proposed definition (which will be discussed below).

The proposed definition satisfies the three adequacy conditions listed above.

(i) The first is almost always trivially satisfied. The traits that are uncontroversially considered to be genetic, and sufficiently well understood at the biochemical level, remain genetic. Usually, these traits were historically identified as potentially genetic because they followed simple Mendelian patterns of inheritance. The first condition in the definition ensures that this must be the case; the second attempts to ensure that, at the phenotypic level, the Mendelian patterns would be recognizable. Sometimes, a trait is called "genetic" when these first two of the definition's three conditions are satisfied while the third is not yet known to be satisfied. This can happen when the allele products of the suspect loci, if any, are not yet known. What the definition given above proposes is that such traits be said to be under "genetic control." They might yet turn out to be genetic but they might not. The alleles at the relevant loci may, either through their immediate or other products, or through a direct regulative role, give rise to Mendelian patterns at the phenotypic level.

(ii) The second adequacy condition is trivially satisfied. Traits could be "biological" without being "genetic" in at least four ways: (a) there might just be a few loci responsible for the trait but it could have low expressivity because of a prominent role of the environment; (b) it could have low expressivity because alleles at other loci could counteract the influence of the alleles that generate the trait (when it is found); (c) the number of loci involved in the ultimate production of the trait could be large[7]; and (d) the immediate products of the alleles responsible for the trait might not be involved in the biochemical characterization of the trait. Moreover, whether or not a "genetic" trait is "inherited" depends on whether the alleles responsible for the trait are also present in those cells that are the precursors for the germinal cells of the organism in question. This might not happen if any of these cells arose through somatic mutations. In organisms such as many "higher animals," which

183

segregate their germ-line, that is, the cells which are precursors to the germinal cells, somatic cell mutations are almost never inherited. In plants and other organisms that do not segregate their germ-lines, such mutations may be inherited. Finally, the differences noted earlier between "genetic" and "inherited" and "heritable" (in the sense of having a high heritability) are all maintained by the proposed definition.

(iii) The situation is a little trickier with the third adequacy condition. As has been emphasized above, the expression of any allele requires interaction with the cellular environment. Without the requisite chemical cues from the cellular environment, no allele would ever be expressed. Now, as already noted above, the constituents of the cellular environment depend at least in part on the (external) environment of the individual organism. The definition proposed above has been framed to identify those traits as "genetic" that turn out to be expressed in as wide a range of (external) environments as possible (to the extent that the organism in question can subsist in that environment). This is what the requirement of high expressivity ensures (and enables the definition to avoid any reference to an unspecifiable "normal" environment). The definition, therefore, illuminates the nature–nurture controversy to the extent that it makes "genetic" traits as independent of the environment, or nurture, as possible. It does not, however, fully underscore a distinction between "nature" and "nurture." That might well not be possible. It is far from clear that the nature–nurture controversy, which keeps on emphasizing nature or nurture rather than worrying about their interaction, is well-conceived. It might well be the case that this adequacy condition can only be partly satisfied not because of any flaw in a proposed definition of "genetic," but because neither genes nor their environment can play the role that disputants about nature and nurture would like them to play.

Beyond the satisfaction of these adequacy conditions, the proposed definition has another advantage, arising mainly because of its third condition. There are situations when "genetic" is used to signify that some alleles or their immediate products, rather than environmental variables, are the appropriate points of intervention if a change in a phenotypic trait is desired. For example, experimental manipulation of organisms is routine in many agricultural contexts. Similarly a desire to alleviate the effects of, or even cure, certain diseases might make such an intervention desirable. The proposed definition captures such usage relatively well. If only a few loci are involved, and their immediate products are part of the chemical characterization of a trait, intervention either at the level of the alleles (through gene manipulation) or at the level of their immediate products may be more efficacious than manipulation of the environment. If a trait is not genetic, for whatever reason, such intervention may only be of dubious value.

There is, however, a potentially very serious problem with the proposed definition. On occasion, traits that are apparently under the control of many loci and usually called "continuous" or "quantitative" (see Chapter 5) are also called "genetic" and "polygenic." The proposed definition excludes such traits from being "genetic." However, this problem may not be as serious as it initially appears because there are at least two reasons not to call the so-called "polygenic" traits "genetic."

(i) It is entirely a matter of assumption that such traits are under the control of a large number of independent loci each making a small contribution to the expression of the trait. Nobody ever produces a list of such loci, though the new strategies mentioned at the end of Chapter 5 (§ 5.7) might eventually lead to such a list. The assumption that quantitative traits can be best explained by reference to many loci (rather than other factors) has been standard since Fisher (1918), and its main role is to show that the normal distribution can be used to capture the distribution of such traits in a given population (see Chapter 5, § 5.2). But what this derivation really requires is that there are a large number of independent factors responsible for the trait; it is irrelevant whether these factors are genetic or environmental. This part of Fisher's argument was no more than an application of the central limit theorem from statistics.

(ii) Such traits seem almost invariably to show a large environmental component of variation (though it is possible that they might not). Note, moreover, that the definition does not exclude, in principle, that a genetic trait might show continuous variation. That depends on the number of loci involved, the degrees of dominance, and the interaction between the alleles. Even a few loci, under suitable conditions, can give rise to a phenotypic trait that varies continuously. What the definition does exclude is a large number of relevant loci and, therefore, excludes those continuous traits that actually are not under the control of a few loci.

But even if the problem with the continuous traits is more serious than is being suggested here, it is still not necessarily a strong argument against the proposed definition. The point here is a philosophical one. A proposed explicit definition of "genetic" is a proposal to move from a "presystematic" to a "systematic" use of the term, a process that constitutes an "explication" of the corresponding concept.[8] Why bother? Presumably only because the presystematic use of the term was leading to difficulties. In the case of "genetic," these difficulties were exactly what motivated this attempt at clarification and even induced others to urge an abandonment of the term

185

altogether.[9] As in many other situations where explications have been offered, the main difficulty with the presystematic use of "genetic" was ambiguity. There is no clear single sense in which the term was being used. Consequently, it is inevitable that some of the senses in which "genetic" was being used, prior to explication, would be lost during that process. That loss, therefore, is not reason enough to abandon the proposed explication, that is, the suggested definition. Thus, even if the prior use of "genetic" to refer to continuous traits was justified, the failure of the definition to capture that use is not, by itself, reason enough to abandon that definition.

It should be apparent, from what has been said so far, that there is a conventional element in explication (and definition). That conventional element is extremely significant: an explication, by being a proposal to establish rules for the use of terms, is a proposal to establish conventions. Thus, the proposed definition of "genetic" is also a proposed convention. Explications, therefore, cannot be judged on what might be called "purely factual" grounds. What are the grounds on which an explications can be judged? Usually, four desiderata are offered: (i) similarity of the explicated concept to the presystematic one; (ii) exactness; (iii) fruitfulness; and (iv) simplicity.[10] While, in general, the requirement of similarity is controversial, in the case of the proposed definition of "genetic" the satisfaction of the definition's first condition, and the first adequacy condition, together ensure significant similarity between the presystematic and explicated concepts. However, the similarity is not complete, as the problems raised by continuous traits show. The simplicity of the definition should be obvious. The question of exactness, as noted above, has been left aside until it is clear that any definition of this sort has a good chance of being adequate on all other grounds. To decide how few loci are few enough, and to decide what level of expressivity is sufficient, will require further conventions to be established.

The most important criterion for the adequacy of an explication, however, is fruitfulness, whether the explication turns out to be useful for subsequent work. Now, it is important to emphasize two factors that are critical in attempts to judge the fruitfulness of any attempt at a definition of "genetic:" (i) the term "genetic" is not a technical term within biology; and (ii) the term has been routinely invoked in contexts that are not "purely" scientific, that is, in social and political contexts. Consequently, the importance of any proposed definition of "genetic" need not necessarily be of any remarkable value in a scientific context but, if it is to be fruitful, at least in one of the contexts, scientific or political, it must make some significant contribution to the debates.

It turns out that in the scientific context the proposed definition is of some value. It helps give a finer classification of many of the traits that are all informally called "genetic." Moreover, it suggests that certain types of research, for instance, heritability analysis, are largely irrelevant to the problem of determining what is genetic. This is consistent with the conclusion reached in Chapter 4 and provides an independent reason to argue against the obsession with heritability. All the same, the fruitfulness of this definition in the scientific context is rather limited because it does not suggest new avenues of research. Abandoning efforts to define "genetic" altogether will have little tangible effect on the pursuit of biology.

Beyond the scientific context, in the political arena the proposed definition, if accepted, would make some important differences. For instance, with respect especially to controversial human behavioral traits, very little of what is popularly claimed to be genetic survives as such. It would cure the habit of calling some trait "genetic" when all that is known is that some unspecified locus has some slight influence in its occurrence – a habit that would lead one to call every trait of every organism "genetic." Moreover, it would emphasize – as most geneticists explicitly do – that the successful modification of these traits through genetic intervention may well be far less plausible than through intervention at environmental levels. It would lead to genetics regaining some of the modesty that it seems to have lost during the last decade.

7.3. CONCLUSIONS

In general, no attempt has been made in this book to decide whether particular traits are genetic or whether – especially for the relatively controversial traits mentioned in Chapter 1 – genetic explanations are forthcoming. Almost all the detailed examples analyzed in this book have been uncontroversial. This has been a book about methods; the application of these methods to the ongoing controversies is being left for a different work. Nevertheless, enough was said in Chapter 4 to discount a genetic reduction for IQ differences, or for any other such phenotypic character for which the only alleged genetic indicator that is available is a high value for (broad) heritability. Enough, also, was said, especially in Chapters 1 and 5, to cast doubt on the potential for genetic reduction of complex human behavioral traits. Nevertheless, it should be emphasized that this book is not intended to denigrate the importance of genetic explanation when it is available or, more importantly, despite what was said in § 7.1, of the intellectual interest and practical value of genetics.

As this book draws to a (perhaps well-deserved) close, suffice it just to note eight points that emerge relatively straightforwardly from the analyses of the previous chapters, but might have been lost in the detail. The first seven of these are points about reduction in the context of genetics. The last is a more general point about reductionism and, technically, draws only on the discussions of Chapters 2 and 3. However, it gains force only because of the detailed application of the general framework developed in that chapter to genetics.

(i) The discussions in Chapter 4 should have conclusively shown that the analysis of heritability, by itself, has little to contribute to genetic explanation. While narrow heritability, **h**, has some value for selective programs in plant and animal breeding, broad heritability, **H**, does not even have such value. To argue from a high value of **H** to ostensible genetic influence is particularly unwarranted because both **h** and **H** are irreducibly population parameters. Luckily, genetics is far from restricted in technique to heritability analysis. That strategy, when based on **H**, is itself of little contemporary value and should be abandoned in favor of the more promising techniques described in Chapter 5 in the pursuit of genetic explanation. (Of course, the analysis of **h** is useful in experimental breeding.)

(ii) As shown in Chapter 5, the type of explanation that emerges in classical genetics requires an abstract characterization of the genotype of an individual as consisting of a hierarchy of loci and alleles. This point was almost universally recognized by classical geneticists, including those who were most systematically pursuing the physical characterization of the genome. The genes obey Mendel's laws for their transmission, as modified by linkage. In particular, classical genetics is an abstract science and does not require any definite physical realization of the genes (alleles or loci) as it pursues the genetic explanation or reduction of traits. This should, after all, come as no surprise. Much of classical genetics was formulated and vigorously pursued before the chemical nature of the gene, or how it acts at the chemical level, was known.

(iii) Traditionally, segregation and linkage analysis have been the most common strategies to be used in the pursuit of genetic reduction and are both potentially useful, though many complications such as variable expressivity, incomplete penetrance, and the existence of phenocopies may both lead to spurious explanations and, sometimes, confound explanatory attempts. These strategies continue to be useful even in the molecular era. However, they only incidentally use concepts that were introduced by molecular biology. A molecular characterization is simply treated as a classical phenotype. For instance, phylogenies are reconstructed using a DNA sequence as the inherited character/trait.

Similarly, linkage analysis can treat such a sequence as a trait known to be genetic (and, in fact, as an allele at a definite locus).

(iv) As the discussions of Chapter 6 show, in sharp contrast, molecular biology and, within it, molecular genetics seek physical explanation or reduction. In this sense, at least, to the extent that reductionist explanations are pursued, molecular biology is similar to ordinary physical study of middle-sized systems. Explanation in molecular genetics, therefore, requires stricter criteria to be satisfied than in classical genetics. Moreover, there is a straightforward sense in which molecular genetics explains and reduces classical genetics, despite the widespread antireductionist consensus among recent philosophers of biology.

(v) Reduction in molecular genetics, and for that matter in the rest of molecular biology, is far from being a completely accomplished task. It has largely proceeded in a piecemeal fashion (far from the dictates of theory reductionism discussed in Chapter 2, § 2.3). However, there are few situations where the prospects for reduction seem bleak (the case of dominance being perhaps the most notable of these). There are no clear anomalies. Most importantly, reductionism continues to generate new lines of inquiry in molecular genetics; this, alone, establishes its importance.

(vi) The physics that lies at the basis for reductionist explanations in molecular biology ("macromolecular physics") is interesting because it is itself barely consistent with, and certainly not (in practice) reducible to, microscopic physics or chemistry. However, one should address this situation within physics where it probably merits more consideration than what has so far been afforded to it. Strictly speaking, it is not a biological problem.

(vii) The different types of reduction offered in classical genetics (leading to "genetic reductionism") and molecular genetics (leading to "physical reductionism") should not be conflated. Both are valuable and scientifically interesting, but they differ both with respect to the substantive assumptions that they make and with respect to the successes they have so far achieved. A conflation of these two types of reduction played an important role in the genesis of the Human Genome Project. Its scientific basis, therefore, remains suspect.

(viii) What this book should have shown is that reduction is a more complicated concept (and procedure) than what even most contemporary reductionists and antireductionists assume.[11] Moreover, the complexities of reductions arise not because of the formal issues raised by those who continue to pursue a relatively analytic orientation in the philosophy of science.[12] Rather, they arise because of substantive issues about the types of representation, the details of interactions, and the approximations that are necessary in the pursuit of scientific explanations (including reductions). There are different types of reduction (at least three, possibly five) that are all conceptually interesting, have exemplars in scientific practice, are actively pursued in many of the natural sciences, and are distinguished on the basis of substantive criteria.

Finally, reduction – in its various types – is scientifically interesting beyond, especially, the formal concerns of most philosophers of science. Reduction is a valuable, sometimes exciting, and occasionally indispensable strategy in science even if it may also fail. There is no reason to doubt that this book has not explored all aspects of reduction in genetics. However, if it has just helped draw attention to the complexities of reduction as philosophers see it while showing how these are related to the actual practice of genetics, it will have achieved some measure of success.

Notes

CHAPTER 1

1. *ESPN*, September 26, 1993.
2. *International Herald Tribune*, September 9, 1992. The remaining examples in this paragraph are also from that source.
3. This argument about the demise of eugenics has been explicitly (and convincingly) developed by Paul (1984). It partly explains why genetic accounts of human behavior continue to emerge periodically in the West.
4. For historical details on the origins of human genetics see, especially, the recent book of Mazumdar (1992). Earlier accounts include Kevles (1986).
5. A potential exception to this generalization are the sociobiologists, who emerged in full force after the 1975 publication of E. O. Wilson's *Sociobiology: The New Synthesis*. However, though the sociobiologists claimed to be drawing their conclusions about human society from the properties of genes, in practice, their genes for various human behaviors remained hypothetically posited entities unknown to the human geneticists. It is, therefore, probably uncontroversial to regard sociobiologists as being outside human genetics. Consequently, the type of reduction to genetics attempted by the sociobiologists will not receive explicit analysis in this book. However, the methodologies discussed in Chapters 4 and 5 are precisely the ones the sociobiologists would have to follow if they were to attempt to ground their analyses on known genetics. Even after that, the success of their project would depend on the epistemological status of those methodologies. Not all of them are equally viable, as Chapters 4 and 5 will show.
6. See Plomin, Owen, and McGuffin (1994).
7. One such conference was held at the Ciba Foundation, London (February 14–16, 1995); another, originally scheduled for 1992, but canceled because of public protests, was held at the Aspen Institute Wye Center, Queenstown, Maryland (September 22–24, 1995). See Editorial (1995) and Bocke and Goode (1996) for more detail.
8. See Koshland (1990), p. 189. Koshland's remarks will be discussed in detail in Chapter 7, § 7.1.

9. For "new genetics" see, e.g., Baron, Endicott, and Ott (1990).
10. The terms "linkage" and "association" used in this Introduction have technical meanings within genetics. They will be properly defined, and analyzed, in Chapter 5. For a rough characterization of "linkage," see § 1.2. The example of bipolar affective disorder is discussed in some detail in Chapter 5, § 5.4. For the original dispute about schizophrenia, see Sherrington et al. (1988) and Kennedy et al. (1988). The most ambitious attempt to date to link schizophrenia to genes is reported by Moises et al. (1995).
11. See Blum et al. (1990).
12. See Pato et al. (1993).
13. Most of the important contributions to this controversy up to the mid-1970s were collected in Block and Dworkin (1976a). At that point the controversy dissipated only to emerge again in the 1990s. Sternberg and Grigorenko (1997) collect together the more recent work. Devlin, Fienberg, Resnick, and Roeder (1997) provide a very critical assessment of the recent claims of a genetic basis for intelligence. To say that there are methodological problems associated with reports that intelligence is genetic (or inherited) is not, of course, to say that it is not. The book will not address the latter claim while endorsing the former (see Chapter 4).
14. The explicit identification of the two disputes goes back to Hogben (1933).
15. For an attempt to elaborate such a strategy, see Gifford (1990); for a response, see Smith (1992).
16. See, for example, Smith (1992, p. 146): "It seems that for any analysis (regardless of the criterion) which reveals genes as the true cause of a trait, there will be a complementary analysis showing the trait to be epigenetically caused . . . [N]o description of a trait as 'genetic' could ever be asserted to be the only (or even the best) appropriate description. If this is truly the case, it is simply not possible to describe traits as 'genetic' in a meaningful sense."
17. The locution "gene-instantiating structure" is necessary to take into account two features: (i) in some unicellular organisms including bacteria, there are circular strands of DNA that encode genes but are not properly called chromosomes; (ii) genes can even be present in mitochondrial DNA segments of higher organisms, and these are not usually called chromosomes. In the rest of this section, this complication will be ignored; it does not substantially alter any of the biology that will be summarized.
18. For the sake of expository simplicity, sex chromosomes are being ignored at this point. In all diploid organisms, the chromosomes divide into two sets, "autosomes," which are paired (and homologous) and special "sex chromosomes," where the two members of the pair are not homologous (in the sense of having the same loci); they usually differ significantly in size.
19. The only exception will be an occasional reference to the sex chromosomes.
20. For a more systematic account of Mendelism, see Chapter 5, § 5.1.
21. Additional complications will arise because of many facts such as: (i) in eukaryotes, genes are not necessarily encoded by contiguous segments of DNA;

(ii) the segments of DNA that encode different genes may overlap; (iii) two chromosomes may exchange parts at a point falling within a segment encoding a gene, etc. A particularly valuable discussion of the vagaries of the term "gene" over several decades is Falk (1986). Some of these complications will be treated in detail in Chapter 6, § 6.5.

22. The term "realm" is being used here instead of the more common "level" so as to avoid any necessary link between "realm" and a "level of organization," which assumes a hierarchical relationship between different levels (see Wimsatt 1976a).

23. See, e.g., Kemeny and Oppenheim (1956), Suppes (1957), Balzer and Sneed (1977), or Balzer and Dawe (1986a,b), all of which do not require reductions to be explanations. This point will be discussed in detail in Chapter 2, § 2.3.

24. In fact, there is no reason to suspect that all the different types of explanation – deterministic or statistical, functional or mechanistic, etc. – can be captured by a single explication.

25. For instance, Wimsatt (1976b) rejects the account of reduction offered by Schaffner (1967b). Part of Wimsatt's concern is about the absence of theories at the level of the explanans. This is, at least partly, a dispute about the nature of explanation – whether explanations must start from well-defined theories. Wimsatt adopts, in his model, a modified version of an account of statistical explanation due to Salmon (1971). Wimsatt's model was criticized by Sarkar (1989) for making deterministic reductions counterintuitive (they were supposed to be statistical explanations of events where one event has probability 1 and all others 0). This, too, is a dispute about the nature of explanation rather than reduction.

26. For a similar characterization of "reductionism," see Schaffner (1993a, p. 411).

27. The motivation for comparing the fate of two individuals in these criteria is to keep determinism distinct from predictability. Determinism is an ontological doctrine, predictability is (arguably) epistemological – at least, like explanation, it has a pragmatic component. Determinism can hold – and the extent to which it holds can often be empirically ascertained – without predictability. One simply sets two appropriately identical systems going and tests whether they remain identical to each other in the future. If so, the parameters by which the initial identity of the systems was determined determines their subsequent behavior. It is not necessary, for this procedure, to have been able to predict what the systems would do from the initial values of these parameters. In the context of physics, this strategy of defining determinism was followed by Earman (1986), and the discussion given here draws heavily on that work. In the genetic context, these definitions are particularly natural since, because of the discrete nature of alleles, two individuals who are (genetically) identical by these criteria can quite easily be found. For the first two criteria, this is trivial: one just looks for individuals with the same alleles (at whatever locus that is of interest). The last two require identical twins. In physics (at least classical physics) because of the continuous nature of many of the relevant parameters (which are needed to specify the initial state of a system), ensuring the identity of two systems to the same level of precision is not quite this easy.

28. See, e.g., Rushton (1995, p. 3): "No behavioral geneticist believes in a 100 percent genetic determinism because it is obvious that physical growth and mental
development require good nutrition, fresh air, and exercise and that children and
neophytes learn best with access to experienced role models." Rushton is one
of the most extreme of genetic reductionists (besides adhering to many highly
controversial and generally dismissed views about the role of race in determining
behavior).

29. Just as genetic reductionism is distinct from physical reductionism, a thesis of
physical determinism is distinct from that of genetic determinism. Physical determinism would simply require that the physical properties of a system determines
all its properties (including its biological properties). It would be characterized
by requiring that two systems with identical initial values for all relevant parameters would have the same behavior (including biological behavior) in the
future [see Earman (1986)]. While this thesis can potentially be investigated in
the context of physical reductionism (Chapter 5), it is not directly relevant to the
concerns of this book and will not be discussed. (This does not mean that it is
not interesting. It is obviously of great interest in the context of neurobiological
determinism with respect to mental phenomena and other similar contexts.)

30. For more general reviews, see Wimsatt (1979) and Sarkar (1992). However, even
these are far from complete.

CHAPTER 2

1. See Stein (1958) for a comprehensive philosophical history of these developments which connects them to the subsequent mid-twentieth century disputes
about the place of reductionist (mechanist) explanations in biology along the
lines sketched in Chapter 6, § 6.1.

2. The most popular logical empiricist thesis of the "unity of science" was simply
that all sciences could be based on observation of the behavior of macroscopic
physical objects; that is, the observations could be described in a "physical language." This was the thesis of "physicalism"; see, for example, Neurath ([1931]
1996) and Carnap (1934). This approach to the unity of science does not call for
reduction. See, also, Chapter 3, § 3.12.

3. Nagel (e.g., 1961) also considered the reduction of one "branch" of science
(that is, an entire scientific domain) to another. For Nagel, this meant that the
theories of that branch were reduced to the theories of the other. All theories were
conceived to consist of a set of laws. In the discussion of Nagel's and similar
views in the text, the reduction of theories will be treated as the reduction of a
set of laws. (Nagel's "branch" plays a role very similar to that played by "realm"
in Chapter 3 and subsequent chapters of this book.)

4. See Stein (1989) for skeptical arguments about the cogency of scientific realism
and its apparent conflict with the type of instrumentalism preferred by most
logical empiricists.

Notes

5. See, especially, Chapter 3, § 3.4 and Chapter 6, § 6.3.

6. Others who have also had to start from a similar point of departure in recent years include Kitcher (1984) and Waters (1990). Note that many of those who reject a special role for theories in reduction, especially Wimsatt (1976b), also explicitly reject logical empiricism. However, by being concerned with formal issues, they follow that tradition at least in part.

7. A similar strategy will be followed with respect to the distinction between ontological and epistemological issues in § 2.2, though the discussion of the ontological issues will be even more cursory than that of the formal issues.

8. For a comprehensive discussion of reduction that emphasizes the formal issues at the expense of substantive ones, see Schaffner (1993a), Chapter 9. That work develops most of the themes that are only touched on in this chapter and provides a very useful guide to the extensive philosophical literature.

9. Nagel's views will merit further consideration in Chapter 3, § 3.7.

10. As some of the discussions of this book will suggest, the linguistic turn was probably more harmful to the analytic philosophers who followed the logical empiricists than to many of the logical empiricists themselves, who were prone to slide into substantive and interesting discussions of science in spite of their avowed preference for merely analyzing the language, rather than the concepts and practices of the sciences (see, e.g., Carnap 1977).

11. For instance, Carnap even conceived of his attempts to evaluate critically the physical concept of entropy as a part of semantics (see Carnap 1977).

12. As Church (1956, p. 65) has pointed out, what Carnap and Tarski had effectively done was to provide a syntactic – therefore, formal – characterization of the truth of a sentence. Semantics, Church insightfully points out, was thus being reduced "in a sense" to syntax.

13. This distinction, therefore, is also not Carnap's (1950) distinction between "internal" and "external" questions. See that work and Carnap (1956) for the distinctions between "syntactic," "semantic," and "pragmatic" as the logical empiricists envisioned them.

14. For an explicit recognition of basically the same distinction, see Hempel (1969). Hempel's "linguistic" issues are the formal epistemological issues discussed here. The disinction is implicit, for instance, in Fodor (1974): token physicalism is an ontological claim about physical determination while the failure of explanation/reduction is an epistemological issue (see § 2.6).

15. This is an assumption, though only a weak one, of a framework-independent realism. Unless the languages at both levels, of explananda and explanans, are at least consistent with each other, this is a potentially inconsistent assumption.

16. In this respect, the point of view endorsed here is the same as that of the logical empiricists. Explanation can still be regarded as a search for causes provided that "cause" is not interpreted ontologically. "Cause" then becomes shorthand for "explanation": to know the cause of some event is to have an explanation for it. From this somewhat deflationary viewpoint, to explain something is to understand it, so long as "understanding" is understood in a relatively public

way; that is, even though an explanation depends on context (see Chapter 3, § 3.1), it is not an individual, potentially idiosyncratic, understanding.

17. Similarly, Fodor (1974), who denies reduction in the mind-body problem, does not deny "token physicalism," which only requires that mental phenomena do not violate physical law. Note that Fodor's position is a rather weak ontological acceptance of physics – all that is demanded is consistency with physics, not even physical determination. For more detail, see § 2.6.

18. This became a standard argument and has been raised many times since (for instance by Kitcher (1984) who argued that these explanations could not satisfy the requirements of Nagel's (1961) model of theory structure and reduction).

19. To note yet another debt to the logical empiricists, this view of ontology is derived from Carnap (1950). See also, Stein (1989).

20. It should come as no surprise that ontological questions have been most systematically pursued in the case of potential neurophysiological reduction (or other such programs to resolve the mind-body question) [see, e.g., Fodor (1974); P. S. Churchland (1986); P. Churchland (1979, 1981, 1984)], where they are obviously of more direct (i.e., scientific) interest.

21. Compare, for instance, Schaffner (1967a,b, 1969, 1976, 1993a) and Hooker (1981a,b,c) versus Wimsatt (1976b, 1995) and Sarkar (1989, 1992).

22. For a useful discussion of other problems of representing laws of nature by universally quantified sentences, see Armstrong (1983).

23. For a similar discussion, see Rosenberg (1985).

24. Of course, anything can be formalized; see § 2.3 above. The point here is that there is no known formalization of these criteria that is more than an exercise in translating them into a chosen formalism.

25. They should not be required to if they originate from positions that deny the importance of reduction in science. Both Hull and Kitcher endorse that point of view, at least with respect to biology. Kitcher (1989), in fact, finds the concept of reduction "outmoded." For a response, see Schaffner (1993a, pp. 499–500). In that work, Schaffner also endorsed the point of view that well-developed laws and theories may not exist in biology.

26. However, only the formal aspects of Nagel's model will be treated here. The substantive aspects will be considered in Chapter 3 (§ 3.8). The version of Nagel's model that will be used is the most detailed (and mature) one from *The Structure of Science* (Nagel 1961). Its basic elements were already elaborated in Nagel (1949).

27. In this sense, explanation is being construed as "deductive-nomological" [see Hempel and Oppenheim (1948)], though the explanandum is not a simple fact, but itself a law. (As noted in the text, theories are construed as a collection of such laws.) "Deduction" will be used here in all such "logical" contexts, where Nagel and other philosophers often use "derivation." The latter term will be used in this book to refer to scientific derivations, which usually involve far less formal procedures than what the philosophers have in mind. [Nagel also periodically uses "logical consequence" instead of "derivation" or "deduction" (e.g., (1961),

p. 352). That distinction will be ignored here.] The terms "reducing theory" and "reduced theory" used here are not Nagel's, who calls the former the "primary science" and the latter the "secondary science."

28. Homogeneous reductions can, therefore, be regarded as a degenerate case of heterogeneous reductions.

29. This is pointed out by Hempel (1969) who, however, sees in Nagel's avoidance of ontological commitment a linguistic emphasis in his treatment of reduction. As noted before, Hempel's distinction is better treated as a distinction between epistemology and ontology. Once that is done, and leaving aside the point that Nagel followed the rest of the analytic philosophers in the linguistic turn, Nagel's emphasis is seen in the more favorable light of being one of epistemology over ontology.

30. Schaffner (1967b) called it the "Nagel-Woodger-Quine" model, noting the similarity of the formalism employed by Nagel and Woodger with that employed by Quine (1964). However, the last work has little relevance to the empirical sciences, and the reference to Quine is therefore being dropped here.

31. In fact, as far as formal models of deterministic explanations are concerned, variants of this model continue to be used to this day.

32. See Leggett (1987) and Sarkar (in press) for further discussion of physical approximations. Wimsatt (1976b) has even argued that deductions are usually used to show how, and to what extent, T_2 does not entail T_1. Approximations will be discussed in more detail in Chapter 3, § 3.4.

33. Popper (1957) endorsed a similar position, which is also consistent with that adopted by Kuhn (1962); see Schaffner (1967b) for an elaboration of these points.

34. See Schaffner (1967b, 1974, 1977, 1993a,b). Variants were developed by Paul Churchland (1979, 1981, 1984) and Patricia S. Churchland (1986).

35. The move from T_1 to T_1' is made in Schaffner (1967b), that from T_2 to T_2' is made in Schaffner (1969). The resulting model is the one discussed in the text. It is embedded in a "Generalized Reduction-Replacement Model" in Schaffner (1977), which is elaborated in Schaffner (1993a,b). Issues connected with replacement rather than reduction are ignored here. Schaffner's (1993a,b) discussion of nonformal questions such as "levels" between which reduction occurs will also be ignored here, but considered in Chapter 3. However, one argument made by him deserves comment here. In defense of his formal model, he argues that the approaches taken by Wimsatt (1976b) and others (e.g., Sarkar 1989, 1992), which refer to mechanisms rather than theories, must be embedded in the deductive framework in order to make sense of the fact that these mechanisms must be fairly "general" [and, in that sense, must be what Sarkar (1992) has called a "fragment of a theory"]. This argument is less than convincing. Restricting attention only to formal issues connected with explanation, because many of these mechanisms are statistical generalizations, they simply cannot be embedded in such a framework: as Salmon (1971) and others have repeatedly pointed out, no simple variant of the deductive–nomological model can

adequately capture statistical explanation. Only deterministic explanations can plausibly be captured by any variant of the deductive–nomological model, and even then some latitude is necessary about what may be considered a theory.

36. Hull (1976) criticizes Schaffner (1976) for a general vagueness about both relations; and Wimsatt (1976b) argues for the necessary context-dependence of "strong analogy." Schaffner (1993a) acknowledges much of these criticisms and follows P. S. Churchland (1986, p. 282) in arguing that T'_1 should be constructed with the analogy in mind; this partially addresses the problem of context-dependence, but not in any formal way. The main point here is that no one seems to have attempted to formalize the relation, let alone done it successfully. Acute context-dependence would suggest the impossibility of finding a general formalism.

37. Suppes's own remarks are tantalizingly brief. Discussing the putative reduction of thermodynamics to statistical mechanics, all he says is (1957, p. 271): "To show in a sharp sense that thermodynamics may be reduced to statistical mechanics, we would need to axiomatize both disciplines by defining appropriate set-theoretic predicates, and then show that given any model T of thermodynamics we may find a model of statistical mechanics on the basis of which we may construct a model isomorphic to T." This gives rise to an obvious objection, that widely different theories (for instance, those of stochastic population genetics and transport phenomena in gases, both of which are based on the diffusion equation) can have isomorphic mathematical structures (see Schaffner 1967b and Sarkar 1989). Later exponents of this approach avoid this objection by requiring – at the nonformal level – that the two theories be about the same domain.

38. In practice Balzer and Dawe generally restrict their attention to the "basic cores" of theories which, following Balzer and Sneed (1977), they believe to be distinguishable from the "special laws."

39. That this condition would be necessary in such schemes seems to have been first noticed by Adams (1959).

40. This is a simpler version of a condition initially introduced by Bourbaki (1968, p. 267).

41. Balzer and Dawe (1986b, p. 185n) attribute this result to S. Feferman.

42. For a detailed discussion of other (internal) problems with the Balzer and Dawe model, see Sarkar (1989, pp. 78–81). For a positive assessment that is at odds with what is said in this paragraph, see Schaffner (1993a, pp. 425–30).

43. However, note that the "semantic interpretation" of theories, championed by van Fraassen (1980), and brought into biology especially by Lloyd (1988), is at least superficially consonant with this approach. It remains to be seen whether an account of reduction based on model–theoretic considerations that escapes these criticisms can be constructed.

44. The focus here is on Wimsatt's model. The general problems – independent of the issue of reduction – with Salmon's model will, therefore, not be discussed. Salmon (1984) discusses these to explain his change of position in the latter work. Schaffner (1993a), Chapter 6, provides a useful summary of the disputes.

45. This distinction, originally due to Nickles (1973), depends on substantive (rather than formal) criteria. It will be incorporated into discussion in the next chapter. Suffice it, here, only to note that for the latter type of reduction Wimsatt expected something like Schaffner's model to hold.

46. Salmon (1984, pp. 36–47) adds two other criteria to the account given here: (i) A is partitioned with respect to the elements of B to define a sample space for the purpose of explanation; and (ii) the marginal probabilities $P(C_j \mid A)$ are also to be computed (to extract some information that would otherwise not be available). These refinements, which were not used by Wimsatt, are of only marginal relevance to the discussion of reduction and will be ignored. The components that are critical in the present context are Salmon's notion of "screening-off" and Wimsatt's notion of "effective screening off"; the refinements do not affect them. More importantly, Salmon (1984) accepts causal explanation as fundamental and attempts to embed the statistical relevance model of explanation within a causal framework. This leads to a different orientation than what was present in Salmon (1971). However, Wimsatt (1976b) simply assumed that statistical relevance showed causal relevance in this model (presumably because of background knowledge) and, in this sense, had already embedded the statistical relevance model in a causal framework.

47. Salmon's attempt to address this problem is less than convincing (1971, p. 80): "In some cases it may be possible to explain a statistical generalization by subsuming it under a higher level generalization; a probability may become an instance of a higher level probability." It is doubtful that the reference classes will continue to be properly definable throughout such a hierarchy. Salmon (1984) does not address this problem.

48. As Glymour (1984) has pointed out, in this case Salmon's account cannot capture a kind of explanation that can quite easily be incorporated into the deductive–nomological model.

49. Note, again, that Wimsatt does not expect his model to be applicable to reductions not involving wholes and parts. For such reductions he expects something similar to Schaffner's model to work. These distinctions will be taken up in the next chapter.

50. In a later version of his model, Salmon (1984) also makes a similar move.

51. Wimsatt uses "is" instead of "consists of" in (c′), which is unnecessarily confusing.

52. This clause is critical to the notion of effective screening-off. According to Wimsatt, it is harder to satisfy than the other clauses: "although upper level descriptions [that is, those that form the basis for C] meeting upper level laws would effectively screen off lower level redescriptions [leading to D], upper level anomalies – upper level descriptions that failed to meet upper level laws – would fail to effectively screen off lower level redescriptions" (p. 704).

53. See, e.g., Carnap (1956).

54. In this sense, as Wimsatt (1976b) and others have pointed out, carrying out a successful reduction adds evidential support for a reducing theory. While

intuitively this may be plausible, whether this idea can be incorporated into any formal scheme of confirmation remains questionable because reductions routinely involve so many approximations that all proposed (competing) reducing theories at any point would probably provide equally successful reductions (see also Chapter 3, § 3.2).

55. Wimsatt followed Schaffner in regarding the connections as synthetic identities. Therefore, the problems noted with "synthetic" and "identity" are also problems for Wimsatt.

56. Wimsatt (1995, p. 228n) accepts many of the criticisms noted in the text though he continues to hope for a liberalized account of identity that would be immune to these criticisms. For other criticisms of the invocation of identities in these contexts see Darden (1991).

57. It is a trivial exercise to formalize these sentences further and deduce (iii) from (ii) and (i'), especially if explanation is to be construed according to the deductive–nomological model.

58. Swanson (1962) seems to have been the first to point out the problematic assumption of deductive completeness in Kemeny and Oppenheim's argument. For Nagel's response, see Nagel (1961), p. 355n.

59. Martin (1972) seems to have been the first to point out explicitly the ontological implications of the assumption that the connections be synthetic identities.

60. See, for example, Schaffner (1967b, 1993a), Sklar (1967), Causey (1972a,b, 1977), Hooker (1981a,b,c), P. S. Churchland (1986), and P. Churchland (1979, 1981, 1984).

61. Moreover, Fodor's justified concern that laws should not have forms such as "(either the irradiation of green plants by sunlight or friction) causes (either carbohydrate synthesis or heat)" can be interpreted purely epistemologically: an explanation invoking such a "law" would not explain. However, the usual situation in putative reductions – including neurophysiological reductions – are more likely to be of the form "either irradiation of green plants by sunlight or light from a plant lamp causes photosynthesis or increase in temperature" at worst. This does explain though, presumably, "photosynthesis or increase in temperature" is not a natural kind; epistemological reduction has no need for natural kinds.

62. A different but complementary objection to viewing the connections as identities has been made by Enç (1976) who has emphasized the epistemological asymmetry between the status of the reducing and reduced theory. Viewing these connections as conditionals is consistent with this asymmetry.

63. A similar point has been made by Hull (1981).

64. Fodor also elaborates a position that was later defended as "token-token" reduction by Kimbrough (1979) in the context of genetics. It does not deny at least partial explanation, but has no role for theories. Effectively it is model that requires explanation of particular genetic phenomena by particular molecular mechanisms. Thus, tokens are connected with tokens, and no types are involved.

65. Note that the claim that supervenience of this sort explicates some notion of determination is controversial. The notion of supervenience that has been given

in the text is basically what Kim (1984, 1987) calls "weak supervenience." This is to be contrasted with stronger notions of "strong" and "global" supervenience that are necessary, according to Kim, to capture the notion of "determination" (or "dependency"). Evidently the notion of determination that is being appealed to is supposed to be stronger than the one used in Chapter 1. Roughly, the stronger notions of supervenience have to be introduced to ensure that the inability of two events that are alike at one level to differ at another level is not due to coincidence. These stronger notions and these distinctions are not particularly relevant in the genetic context, and will not be pursued here. In any case, Kim (1989, p. 40) has conceded that "if a relation is weak enough to be nonreductive, it tends to be too weak to serve as a dependence relation; conversely when a relation is strong enough to give us dependence, it tends to be too strong – strong enough to imply reducibility." Moreover, why weak supervenience should be regarded as an insufficient concept of determinism is far from clear. Not only does it capture the type of determinism described in Chapter 1, but there is also a potentially weaker concept of supervenience that would not violate most intuitions about determination and would simply require that two entities from the realm of the potential explanandum can be different only if the relevant entities in the realm of the potential explanans are different, but routinely not the converse.

CHAPTER 3

1. See, e.g., Kauffman (1972) for a cybernetic representation of a cell.
2. By a suitable choice of predicates, any derivation, including those that involve approximations, can be cast in the form of a logical deduction. However, there are few if any known cases where this process generates insight or clarity.
3. It should be emphasized that explanations of this sort are not to be regarded as inferior to those – typically in physical contexts and in ecological and evolutionary theory in biology – where the derivations involve the systematic use of mathematical reasoning. There is no a priori reason to assume that all domains of science require mathematical representation or reasoning. Molecular biology provides, perhaps, the most obvious examples of blatantly nonmathematical but nevertheless successful and rigorous science.
4. In some weak and intuitive sense, these factors are the ones that are responsible for what is being explained (otherwise, they would not be explanatory). It should not be assumed that the factors must provide necessary or sufficient conditions for whatever it is that is being explained. Certainly, they cannot be taken to determine the feature being explained (in the sense of "determine" used in Chapter 1). However, enough detail can presumably be filled in to get a sufficient story. Those who prefer to use ontological terms would presumably want to call these explanatory factors "causes."
5. It is possible to avoid using "factors" altogether and use "explanans" instead, with "explanans" now no longer being understood as a linguistic entity. However,

it seems better to introduce "factors" both because that term is often used in the same way in ordinary scientific contexts and also to avoid the misleading traditional connotations of "explanans."

6. In Sarkar (1996) this criterion was called "hierarchical structure."
7. In Sarkar (1996) this was called "spatial instantiation."
8. The meaning of "approximate satisfaction" will be spelled out in greater detail in § 3.3.
9. This point was independently made by Sarkar (1996) and Thaler (1996).
10. Since criterion (iii) depends on (ii), that is, (iii) cannot be satisfied unless (ii) is, there are only five possible types (a type that involves the satisfaction of (iii) alone, or only (i) and (iii), is precluded). This argument assumes that (i) must at least be approximately satisfied, but not necessarily fully satisfied. [If (i) is not satisfied at all, then this would generate more possibilities. However, this is not a relevant option in the context of reductions; see § 3.3.]
11. However, for an interesting attempt to find such reductions in science, especially physical chemistry, see Ramsey (1995). These reductions will not be considered any further in this book, which confines itself to reduction as explanation.
12. However, Wimsatt's concerns are primarily ontological. This makes him less inclined to recognize the importance of nonspatial hierarchies in explanation.
13. Thus, "realm" corresponds to Nagel's "branch"; the "F-realm" corresponds to his "reducing branch." Reference is being made to the rules operative in different realms rather than to laws to emphasize the point that the account of reduction being given here gives no special status to laws and theories.
14. The concept of F-justification replaces that of "physical warrant" in Sarkar (1989, 1992, 1996).
15. This point has been particularly forcefully developed by Cartwright (1983).
16. See, e.g., Barr (1974) and Causey (1977). Cartwright (1983, p. 104) also makes the point that is made here in the text.
17. Causey (1977, p. 23) seems to express a customary point of view when he simply states "[o]f course, it would be nearly impossible to try to characterize all kinds of allowable approximations" as if that were reason enough to ignore the subtleties of approximations altogether.
18. Mathematically justifiable corrigible approximations (which, in general, should be ones whose effects are estimable) correspond more or less to what Laymon (1991) calls "transformational" approximations. Mathematically justifiable incorrigible approximations correspond quite closely to his "calculational" approximations.
19. This example will be discussed in detail in Chapter 6, § 6.2 – § 6.3.
20. Counterfactual approximations have been called statements of "ideal conditions" by Barr (1971). [That Barr's analysis was actually one of approximation, rather than idealization, was pointed out by Schwartz (1977).]
21. Cartwright (1983) has argued that in explanation (in general), the explanandum is often very well-confirmed and, therefore, should be taken as true. However, the explanans ("fundamental theory") explains the explanandum only through

counterfactural (and, perhaps, also otherwise epistemologically dubious) approximations. Therefore, the explanandum is false. This position is slightly more extreme than the one adopted here. Translated into the present context, Cartwright's position would be that when counterfactual approximations are used for a reduction, the reducing rules are incorrect, rather than that the reduction is problematic (which is the view advocated here). However, if the reduced rules, phenomena, etc. are taken to be part of the evidence that should support the reducing rules [as Wimsatt (1976b) suggests], then, indeed, a failure of reduction indicates problems with the reducing rules. This comes close to Cartwrights position.

22. See Schaffner (1993a) and even Wimsatt (1976b). Wimsatt's account of strong ("inter-level") reduction makes reductions transitive because of its use of Salmon's model of explanation. For weak [type (a)] reductions (which he calls "intra-level"), Wimsatt explicitly suggested that they could be intransitive.

23. In Wimsatt's (1981) terminology, models using approximations of this sort would be "robust."

24. In Wimsatt's (1976b) model of reduction, the refinement of F-rules would be a case where an attempted reduction leads the reduced theory to correct the reducing one. However, there is nothing in that account that corresponds to the invention of new F-realms, since the levels of organization remain fixed. Other accounts of reduction ignore both these possibilities [but see the discussion of Nagel (1961) in § 3.11].

25. Sarkar (in press) has argued that Einstein's (1905) paper on Brownian motion is an exemplar of this type of insight.

26. In general, differentiating levels provides an easy way of differentiating realms when the hierarchy condition holds. In such a situation "level" will be used identically with "realm."

27. For a similar earlier use of directed graphs, see Wimsatt (1976a).

28. However, there would be no harm in calling the levels of an abstract (nonspatial) hierarchy "levels of organization." This is sometimes done [see, e.g., Wimsatt (1976a)].

29. See, e.g., Darden and Maull (1977) and Schaffner (1993a).

30. Note that it is not being suggested here that all explanations in molecular biology are reductionist of either type (d) or (e). Attention is simply being restricted to those explanations that appear to be "parts-whole" explanations, that is, at least of type (d). [For an extended discussion of information-based nonreductionist explanations in molecular biology, see Sarkar (1996). Sarkar (1988, 1991) discusses functional explanations in molecular biology that are not directly reductionist.]

31. Cartwright (1983) has correctly pointed out the intriguing fact that a similar conclusion holds in many other situations where quantum mechanics is presumed to be the fundamental theory.

32. This point was explicitly incorporated by Wimsatt (1976b) and Sarkar (1989). It was also implicitly accepted by Schaffner (1967a).

33. This point has been previously noted by Sklar (1967), Nickles (1973), and Wimsatt (1976a,b).
34. The examples given in this chapter show that these are almost all weak reductions [that is, of type (a)].
35. Note that this position should not be confused with one demanding ontological elimination following a replacement as, for instance, advocated by P. Churchland (1979, 1981, 1984) in the context of the relation between folk psychology and neurobiology. Hull (1972, 1974) endorsed a similar kind of ontological elimination (implicitly) and replacement (explicitly) in the context of the relation betweeen classical and molecular genetics.
36. Details of this point will be discussed in Chapter 6, § 6.3.
37. For the development of the only systematic interpretation of biological research along these lines, see Haldane (1939).
38. However, Wimsatt (1995) claims to deny ontological eliminativism on ontological grounds in the sense that he maintains a sort of realism for entities, properties, and processes at all levels of organization, although endorsing a liberalized, presumably both epistemological and ontological, reductionism. Wimsatt's realism is admittedly "local" – "reality" is to be ascribed to entities, properties, processes, and so forth, on grounds of "robustness," by which he means a sort of triangulation to a particular point through a variety of diverse methods [see Wimsatt (1981)]. Robustness, according to this scheme, points to the reality of things at different levels of organization. However, Wimsatt's realism is a rather innocuous doctrine. There is no strict ontological reduction and the entire discussion, with its valuable analyses of types of scientific strategy, is easily recast with talk of ontology thrust aside. For instance, robustness seems, in practice, only to add to the evidential support for a particular claim, and can be recast as a generalization of the traditional emphasis on the variety of evidence (in confirmation theory). It is at least arguable that Wimsatt's ontological excursions are extraneous to, and detract from, the value of his epistemological analysis of reduction.
39. This episode is treated in philosophical detail by Stein (1958) and Sarkar (1989).
40. The qualification "usually" is necessary in this sentence only because of the possibility that nonexplanatory correlations can be used for prediction. See Chapter 1, § 1.4.
41. See Wimsatt (1976a,b), Sarkar (1989), and Darden (1991).
42. Note that, whereas the different types of reduction that have been distinguished here on the basis of substantive criteria are neutral to the form of a reduction (insofar as the question is that of the nature of reduction), they do not remain similarly neutral when the question is that of research strategy. The fact that each of these types subsumes research strategies that are explicitly followed in genetics is, therefore, an argument in favor of this analysis over the traditional ones (though this argument is extraneous to the direct concerns of this book).

Notes

43. See Delbrück (1949) and the recollections of Stent (1968), and the discussion in Chapter 6, § 6.1. Stein (1958) and Sarkar (1989) have put this program into its proper philosophical perspective.
44. For a detailed analysis of this reseach strategy, see Wimsatt (1976b) and Darden (1991). Wimsatt's piece is more directly concerned with reduction; Darden's is more general. Wimsatt insists on the necessity of synthetic identities (connecting entities in the realm being reduced and the **F**-realm) in this process but, as was argued in Chapter 2, § 2.5, all the identification that is necessary for this research strategy to be pursued can be captured by (synthetic) conditionals. For historical details on the Morgan group, see Carlson (1966), Allen (1979), and Kohler (1994). The last work is more sociological in its orientation and, therefore, less directly concerned with the issues discussed in this book. Nevertheless, it provides the most complete history of the Morgan school to this date.
45. See Nagel (1961), p. 358; all other quotations from Nagel in this section are from this source.
46. This defense of reductionism has also been previously urged by Waters (1990), one of the few works that pay any attention to Nagel's nonformal considerations. Sarkar (1989) also argued for the importance of that part of Nagel's analysis.
47. Elsewhere in the same paper, however, Quine claims that he is doing no more than endorsing supervenience [as defined by Davidson (1970)]: "It is not a reductionist doctrine of the sort sometimes imagined. It is not a utopian dream of our being able to specify all mental events in physiological or microbiological terms. It is not a claim that such correlations even exist, in general, to be discovered; the groupings of events in mentalistic terms need not stand in any systematic relation to biological groupings. What it does say about the life of the mind is that there is no mental difference without a physical difference" ([1977] 1979, p. 163). How this is supposed to cohere with what was quoted in the text remains somewhat mysterious. Here, as elsewhere, one should not accuse Quine of consistency and of that smallness of mind that goes with it [see Harding (1975) and for more details on the last point].
48. The point that should be emphasized is that the reduction to a physical language (or a phenomenal language) involves a different concept of "reduction" than the one being used in this book. Among other things, it involves no explanation (see, also, Causey (1977, pp. 106–7)]. For an account of this other kind of reduction, see Bonevac (1982).
49. The second version of the thesis is the one that gained most currency. See Chapter 2.
50. In fact, what is surprising is the persistence of the belief that reductions can routinely lead to unification; see, e.g., Causey (1977), Maull (1977), Hooker (1981c). (Maull's paper is ostensibly an attempt to show how science may be unified without reduction. However, by "reduction" she means Nagel- and Schaffner-type models. The alternatives she presents qualify as strong reductions as construed here.)

51. See Sarkar (1988, 1989, 1991) for a discussion of functional explanations in molecular biology.

CHAPTER 4

1. The various concepts of heritability will be exactly defined in § 4.2. What is said at this point is independent of the differences between these concepts.
2. For the "sordid" history of research on the heritability of IQ, see, for example, Kamin (1974), Block and Dworkin (1976b), Fancher (1985), Tucker (1994) and Gould (1996). Criticisms include accusations of various forms of racism from the 1920s to the 1990s as well as claims of outright fraud on the part of one of the most prominent IQ researchers (Cyril Burt). For the other traits mentioned here, see Bouchard et al. (1990) and Bouchard (1994). It is, of course, open to question whether these characteristics are stable enough to be designated "traits." As noted in Chapter 1, § 1.2, there is no technical definition of that term. Hence, they are called putative traits here to record potential problems with this designation.
3. A third concept of heritability was distinguished by Jacquard (1983). It has found almost no use so far and will not be discussed here.
4. See Bell (1977) for a pioneering but, unfortunately, rather sketchy history. It seems to be the only history of the introduction and use of "heritability" that exists.
5. In most discussions, for either of these criteria the locution "can be accounted for from a genetic basis" is usually replaced by "is determined by the genetic basis (or the genes)." However, the latter formulation conveys no more information than the former and has the potential to be misconstrued as some thesis about genetic determinism that many proponents of the use of heritability would not endorse [see, e.g., Plomin, DeFries, and McClearn (1990)].
6. Note that the use of "reducible" and "irreducible" in this context should not be confused with the use of "reduce," "reductionist," and so forth that is the general concern of this book (though there are straightforward analogies).
7. A weaker requirement would be that the influence of other individuals be small. However, as Figure 4.4.4 will show, narrow heritability can depend on the frequency of alleles in a population in such a way that it can take any value between 0 and 1 (for a trait whose origin can entirely be attributed to alleles at a single locus). Therefore, even a weaker notion of reducibility would be violated by heritability; it will not be explored any further. It should be emphasized that since this book and, especially, this chapter is concerned with conceptual issues, it is the stronger notion of reducibility that is important.
8. Putting it in another way, what is and what is not genetic should depend entirely on what an individual's genotype is, and how it interacts with its (physical) environment (especially during development). It should not depend on the properties – genotypic or otherwise – of other individuals.

9. Precisely because of this, the ambiguity in the requirements (as to which concept of heritability is being invoked) is harmless. However, the claim that no structure is being asumed for the genotype is only true for broad heritability (**H**).

10. Critics of the use of heritability have not sufficiently emphasized this point, perhaps because most of them not only doubt the value of heritability analysis but also doubt that the controversial traits (such as IQ) for which it has been used have any significant genetic origin. However, even those who continue to advocate the continued use of heritability analysis sometimes do not seem to recognize this point sufficiently [for example, Khoury, Beaty, and Cohen (1993)].

11. See, e.g., Layzer (1974), Kempthorne (1978), Jacquard (1983), Falconer (1989), Wahlsten (1990), and Khoury, Beaty, and Cohen (1993) for discussions of estimation problems. These will only be briefly noted in § 4.5 but not discussed in any detail in this book.

12. The use of "2" in the exponent does little more than reflect the fact that a variance is the square of a standard deviation! It was probably psychologically reassuring in an era when standard deviations were common fare while variances were relatively new.

13. See, for example, Lewontin (1974) or Feldman and Lewontin (1975).

14. For example, Plomin (1994, pp. 43–4) states: "[**H**] is a descriptive statistic that estimates the proportion of phenotypic variance . . . that can be accounted for by genetic variance."

15. Most of this is well-known. It was implicit in the work of Hogben (1933) as early as 1933 and was particularly clearly articulated by Lewontin (1974).

16. This caveat is important and should not be confused with a somewhat related, but nevertheless different point, that (classical) genetics was the study of "gene differences" as observed through phenotypic differences [see, e.g., Haldane (1936b)]. The point there was simply that in the premolecular era of genetics allelic differences could only be experimentally inferred from the observed phenotypic differences and their transmission. Nevertheless, it was assumed that the alleles were precise units (to be identified or measured in their own units). The epistemological limitation was one of technology, not principle, in contrast to the situation in the case of **H** (or **h**).

17. Such a view is implicit, for instance, in Plomin (1994). Well before **H** was explicitly introduced, this use of the analysis of variance was noted and very carefully analyzed by Hogben (1933).

18. This is not to suggest that the biological–nonbiological distinction is particularly clear. Perhaps it is not clear because the nature–nurture distinction is far from clear, as the discussions of this chapter will show [see also Hogben (1933)]. In contrast, the genetic–biological distinction is a little more straightforward; see Chapter 7, § 7.1.

19. For an explicit example of such a claim, see Bouchard et al. (1990).

20. On the other hand, the use of **h** does implicitly involve an abstract (though trivial) hierarchy assumption since it is assumed that the genome has a structure

consisting of several loci, and each locus subsumes several alleles. On this point, see the discussion of Fisher's reduction of biometry to Mendelism in the next chapter.

21. Formally, these possibilities are: $\mathbf{Cv}(G, E) \neq 0$ [that is, Equation (4.A1.8) fails] and $\mathbf{V}(R) \neq 0$ [that is, Equation (4.A1.13) fails].

22. Note that if there is a negative statistical correlation between genotype and environment, that is, $\mathbf{Cv}(G, E) < 0$, then it is trivially true that $\mathbf{H}$ overestimates $\mathbf{V}(G)/(\mathbf{V}(G) + \mathbf{V}(E))$.

23. Schmalhausen (1949) has even argued that this property is itself inherited by individuals. In those cases where this is true – in an ironic twist to the debate – even the minimal interpretation of heritability and, therefore, any potential use of the concepts will be invalidated because of a "genuine" feature of inheritance.

24. The same argument has also been very clearly articulated by Lewontin (1970, 1974), apparently independently, and picked up by many others [e.g., Block and Dworkin (1976b)].

25. The argument given here does not require that the norms of reaction must be horizontal. That simplification is only assumed for didactic purposes.

26. The analysis given below will also clarify exactly what $\mathbf{V}_a$ and $\mathbf{V}_d$ are, a point that was left partly unresolved in § 4.1.

27. For other models illustrating the same point, see Roughgarden (1979, pp. 145–64).

28. $\mathbf{H}$ can also be estimated through what is only apparently a conceptually different technique, namely, *path analysis*, a method that goes back to Wright (1921). However, it can be demonstrated that the two methods, path analysis and ANOVA, are equivalent in the sense that they permit the estimation of $\mathbf{H}$ in exactly the same circumstaces. Thus, the objections noted in the text are as applicable to path analysis as to the methods mentioned there. (Moreover, even if path analysis truly did provide an alternative way to estimate $\mathbf{H}$, none of the interpretive problems noted in § 4.4 would in any way be ameliorated.) For a different (and important) objection to this use of path analysis, namely, the underdetermination of $\mathbf{H}$ values by path diagrams with radically different interpretations, see Fine (1990).

29. Among geneticists, at least, this is well-known; see, e.g., Stern (1973), Kempthorne (1978), Jacquard (1983), Falconer (1989, pp. 173–6) and Khoury, Beaty, and Cohen (1993, Chapter 7). Layzer (1974) has given an explicit discussion of how, if Equation (4.1.3) does not hold, any attempt to use twin studies to estimate $\mathbf{H}$ involves an attempt to estimate more parameters than there are relations among the measured variables.

30. This is reported by Mann (1994).

31. See Plomin and Loehlin (1989).

32. A detailed history of human genetics is yet to be written. For Hogben and Haldane, see Mazumdar (1992) and Kevles (1986). The latter work takes the story into the 1950s, by which period human genetics had finally emerged from its

eugenic roots – thanks largely to the efforts of Lionel Penrose – and, arguably, taken its modern form.

33. That the alleged heritability of IQ is controversial is well-known; see, e.g., Kamin (1974), Block and Dworkin (1976b), Fancher (1985), and Gould (1996). What often does not receive sufficient attention is that the relation between IQ and what may be intuitively called general cognitive ability is less than certain. For a particularly incisive discussion of this issue, see Block and Dworkin (1976b). It should come as no surprise that the recent collection, *What is Intelligence?* (Khalfa 1994), has only one cursory reference to IQ. Moreover, recent work on intelligence and cognitive ability (in general) has tended not only to be less focused on IQ, but also less reliant on heritability analysis than before [see Sternberg and Grigorenko (1997) and Devlin et al. (1997)]. In particular, Plomin (1997) has strongly urged and begun experimental work based on methods that will be discussed in the next chapter. This seems much more promising.

34. See, for example, Jensen (1969, 1973); Herrnstein (1971, 1973); Shockley (1972) and Herrnstein and Murray (1994).

35. See, e.g., Plomin, DeFries, and McClearn (1990), pp. 367–8.

36. For alternative equivalent – though less explicit – accounts, see Feldman and Lewontin (1975) and Jacquard (1983).

37. The terminology followed in this section, which distinguishes the genotype–environment covariance from the genotype–environment interaction, is common [see, e.g., Layzer (1974) or Plomin, DeFries, and McClearn (1990)] but not universal [see Fancher (1985) or Wahlsten (1990)]. Both the genotype–environment covariance and the interaction would generate statistical correlations between genotypic and environmental variables. In the discusssions ot the text, "correlation" will be used to subsume both. Note that the notation for the variance and the covariance used in this section is slightly more explicit than what will be used in the rest of this chapter; it is needed here for clarity.

38. Jacquard (1983) and others (see, e.g., Plomin, DeFries, and McClearn (1990), pp. 223–4) have argued that it is also possible to define R in such a way that Equation (4.3.9) is true. This procedure apparently involves the requirement that R be as small as possible (using a least-squares fit), compared with G and E, that is compatible with the data. (The procedure to be followed would presumably be analogous to the one used to define $\mathbf{V}_a$ in the model analyzed in Appendix 4.2.) Moreover, there is no theoretical reason to suppose that R defined in such a way will be the same as the one defined in the text, or that G and E will have the interpretations given there (see Dudley 1990).

39. Layzer (1974) also proves a rather remarkable result: if Equation (4.A1.8) holds, the expectation value of R^2 is a minimum.

40. Technically, all that is required is that $R(\vec{g}, \vec{e})$ be constant. However, since only variances are involved in the definitions of $\mathbf{H}$ (or $\mathbf{h}$), by an appropriate choice of origin, the constant can be set to 0.

41. This and the problems mentioned in the previous paragraph have sometimes led to the claim that $\mathbf{H}$ cannot be "defined" in such a circumstance [see, e.g., Moran

(1973)]. However, the problem seems to be one of interpretation rather than definition.

42. This point is particularly important in the context of debates about the alleged immutability of IQ scores because **H** is said to be about $0 \cdot 7$. There is no scientific basis for this claim.

CHAPTER 5

1. From the point of view of statistical inference, this model serves as the null model. For a discussion of methodological problems with these methods from the point of view of potential errors in statistical inference see, for instance, Khoury, Beaty, and Cohen (1993, Chapters 8 and 9). Of the methods discussed in this chapter, the allele-sharing method is nonparametric whereas the others (including linkage analysis in practice) are parametric. Issues connected with the differences between parametric and nonparametric methods are not germane to reductionism and, as with other issues connected with statistical methodology, will be ignored in this book except occasionally in the footnotes. These issues deserve much more philosophical attention than has so far been afforded to them.

2. In § 5.6 it will be pointed out that the type of reduction that results from the techniques discussed in this chapter is the abstract hierarchical [type (c) in the classification of Chapter 3, § 3.2]. However, "genetic reduction" and "genetic reductionism" will be preferentially used throughout this chapter and in the rest of this book because of their obvious didactic advantage.

3. The modifications introduced by polyploidy (or for that matter haploidy) are technical rather than conceptual. The general conclusions about the various methods, and especially about the status of reductionism, reached in this chapter are not affected by ploidy differences.

4. See, for example, Olby (1979, 1985), Monaghan and Corcos (1990), and Falk and Sarkar (1991) for discussions of this issue.

5. From the point of view of the conceptual structure of explanations in classical genetics, any reference to the physical realization of loci and alleles is not necessary. This point will be discussed in detail in § 5.6. However, the physical basis of inheritance is being invoked here in the interest of presenting as clear an account of these laws as possible.

6. See, for example, Galton (1889) and Pearson (1893, 1900).

7. See Stigler (1986), Chapters 8, 9, and 10, for a detailed history of statistical developments. Provine (1971) discusses some of the biological issues. Provine emphasizes that the heat of the dispute was to a large extent due to personal antagonism between Weldon and Bateson.

8. For a detailed history, see Froggatt and Nevin (1971) as well as Provine (1971).

9. This aberrant historiography goes back to Huxley (1942). It was implicitly endorsed by Provine (1971). For an account of an "evolutionary synthesis" that

allegedly occurred after 1930, rather than from the work of Fisher, Haldane, and Wright, see Mayr and Provine (1980). This revisionist historiography is wildly implausible but a discussion of this issue is beyond the scope of this book.

10. For a detailed history of this law, though one without much technical detail, and of the controversies surrounding it after the rediscovery of Mendelism, see Froggatt and Nevin (1971). Provine (1971) also discusses the law but provides even less detail.

11. Mid-parental averaging is necessary for those traits for which the value in the two sexes is systematically different (that is, there is sexual dimorphism). For instance, women are systematically shorter than men in the same population. Mid-parental averaging for height would involve multiplying all observed values of women's heights by a factor (>1) to correct for the systematic difference. This factor was obtained empirically, and the justification for the entire procedure was ultimately empirical [see Stigler (1986), Chapter 8].

12. See Moran and Smith (1966) for more details on this point and for general critical annotations of other arguments in Fisher's notoriously difficult paper.

13. Waddington's remark was part of a polemic directed against mathematical population genetics. According to him, its only other achievement was that it showed that "ordinary, well-known Mendelian genes could respond to the processes of natural selection and could thus be entities on whose variation evolution depends" (1957, p. 61). Thus, mathematical population genetics is of little explanatory value for the actual details of evolution. For a response to Waddington, see Provine (1986).

14. In the context of the relationship between microscopic and macroscopic physics, the same point (about only establishing consistency) has been made by Leggett (1987).

15. This probably partly explains why heritability analysis arose in the first place (in quantitative genetics), and why it gained as much popularity as it did in spite of the problems noted in the previous chapter.

16. This statement is ahistorical to the extent that the phenotype–genotype distinction was only explicitly made by Johanssen (1909). However, the distinction was implicit in both Mendel and Galton. At best, Johannsen's importance lies in reviving the distinction in a Mendelian context where, during the initial phase of the recovery of Mendel's work, it had largely been lost as characters and factors were routinely conflated with each other.

17. No doubt, Pearson's explicit positivism encouraged this skepticism. Norton (1975) has argued for this interpretation.

18. For a particularly clear treatment, see Thompson (1986).

19. Mazumdar (1992) has even argued that the liberation of human genetics from its eugenic antecedents required the replacement of semiintuitive analysis of pedigrees by more reliable quantitative methods. According to her, this occurred in the 1930s, mainly due to the efforts of Hogben and Haldane.

20. In the same paper Bateson also suggested, though without similar quantitative considerations, that albinism and PKU were recessive traits controlled at one

locus. Finally, he described the sex-linked inheritance of hemophilia and color-blindness but provided no explanation.

21. X-linked traits are those whose loci are on the X-chromosome. In human males there is only one X-chromosome. In females there are two. Therefore, a recessive X-linked trait is always phenotypically distinguishable in males, but only in females who are homozygous for the allele that is involved.

22. Morgan et al. (1915, p. 5) attribute this discovery to Bateson and Punnett in 1906. However, as Darden (1991, p. 121) has pointed out, Correns (1900) and Castle (1903) had already noted the violations.

23. For a history of these developments, see Carlson (1966) and Darden (1991).

24. For a short and clear discussion of these developments, see Ott (1991), pp. 14–19. Wimsatt (1992) discusses the philosophical issues connected with the early use of mapping functions.

25. Since linkage analysis only directly establishes relative positions, the physical location of at least one locus is necessary for this step. Determining these physical locations is called "physical mapping." In the interest of expository simplicity, this complication will be ignored here.

26. This is particularly true when the problems discussed in the next section (§ 5.5) must be accosted.

27. For an introduction to the various methods, see Ott (1991). Straightforward linkage mapping is a parametric method since a definite model (linkage to a known locus with its mode of inheritance) is tested. For an example of linkage detection that involves a nonparametric method (and is not linkage analysis as defined here), see the discussion of the allele-sharing method in § 5.7.

28. Wimsatt (1992) seems to be the only philosophical treatment of the problems of linkage mapping. It is restricted in scope to the period before 1930 and contrasts the approaches of linkage mapping through linkage functions (for instance, by Haldane) and the semiempirical approach of the Morgan school.

29. This amounts to the selection of a model to be tested. This is exactly why linkage analysis is a parametric method even though it is based on a violation of a model in the sense that it uses violations of the law of independent assortment.

30. For a discussion of multiple loci see, e.g., Ott (1991), Chapter 6.

31. The terminology "*cis-*" and "*trans-*," introduced by Haldane (1942) from a chemical analogy, reflect nothing more than the necessity of distinguishing the two cases (since the subscripts are no more than conventions, that is, B_1 could just as well have been labeled B_2).

32. See Egeland et al. (1987).

33. See Baron, Endicott, and Ott (1990) for an early discussion. Baron et al. (1993) report the diminished support for linkage from three Israeli pedigrees.

34. See Kelsoe et al. (1989).

35. See Botstein et al. (1980).

36. See Weatherall (1991), pp. 117–18 for a discussion of RFLP mapping. The examples given in this paragraph are all from that source.

37. See Cook-Deegan (1994) for a brief history of research on Huntington's disease.

38. If, instead, the reconstruction of geneologies is the aim, what has to be computed is the probability of the observed phenotypic distribution, given the hypothesized geneology; the known genetic model of inheritance; and the assumed relationship between genotype and phenotype [see Thompson (1986), Chapter 4].

39. Because it is not particularly relevant to reduction, the discussion given above glossed over all quantitative questions connected with such inference. The examples chosen were particularly simple and, as in the case of the pedigrees discussed by Bateson and Haldane, the claim that there was quantitative agreement between the data and the theoretical expectations was left intuitive.

40. See Thompson (1986), Chapter 8 for a review.

41. In the absence of recombination, multiple loci can be treated as a single locus with more alleles. For instance, if there are two loci with alleles A_1, A_2 and B_1, B_2 at them, it could be treated as a single locus with alleles $A_1B_1, A_1B_2, A_2B_1, A_2B_2$.

42. For details, see Ott (1991), Chapter 6.

43. This is penetrance in a broad sense. In some medical contexts, it is distinguished from penetrance in a narrow sense, which is defined as the conditional probability of being affected with a given disease, given a genotype. If the phenotype is identified with being so affected, the two definitions coincide. The narrow sense of penetrance is apparently only used in such medical contexts [see Ott (1991), pp. 147–8]. Note that penetrance is sometimes treated simply – and misleadingly – as a property of a genotype rather than as a relational property (e.g., by Lander and Schork (1994)].

44. The standard cases of phenocopies are those that can be routinely introduced by heat treatment in insects including various species of Papilio and Drosophila [see Goldschmidt (1938)].

45. See Humphries, Kenna, and Farrer (1992) and Lander and Schork (1994).

46. The restriction to two alleles (for each given locus) is only because of diploidy. Haploidy would allow only one allele per locus, polyploidy requires more than two. Neither of these cases alters the argument in the text. However, polyploidy introduces significant technical difficulties. These are far worse in the case of linkage analysis than in the case of segregation analysis. Some, but not much, progress has been made in either case. This is probably because genetic studies concentrate on micro-organisms and animals rather than plants. Plants are more likely to be polyploid.

47. This somewhat surprising observation was noted – independently – by Sarkar (1996) and Thaler (1996) apparently for the first time in a historical context. As the discussion in the text will show, it was quite generally recognized by the classical geneticists.

48. In the eukaryotic genome, however, the correspondence is not exact. For instance, it is now known that loci/genes can overlap, and that there can be loci/genes within others (see Chapter 6, § 6.5). Traditional linkage analysis cannot lead to the possibility of such a nested but "linear" order.

49. See Morgan, Bridges, and Sturtevant (1925), p. 88.

50. See Haldane (1939). Haldane's prediction of the failure of mechanism was based on his espousal of dialectical materialism during this period. In the late 1950s Delbrück also suggested a failure of linear correspondence between loci as revealed by linkage analysis and physical positions on chromosomes; in his case, the hope was that it would finally demonstrate the necessity of "complementarity" in molecular biology [see Fischer and Lipson (1988)].

51. Note that this is true even when there is genetic heterogeneity. In that situation, while alleles at many different loci lead to the same phenotype, in each individual case a single locus or very few loci are involved.

52. This section draws heavily on the review by Lander and Schork (1994), though it does not agree with all their conclusions.

53. For example, see Plomin (1994).

54. In the case of segregation or linkage analysis, the identity of alleles by state but not by descent would generally decrease the statistical significance of an inference.

55. See Hamer et al. (1993) for the original study. Hu et al. (1995) report the follow-up study. Byne (1994) was probably the first to point out the caveat mentioned in the text.

56. See Bailey (1995). Bailey provides a useful summary of the dispute to date and concludes that the data do not allow any definitive conclusion to be drawn.

57. See Marshall (1995). See, also, Holden (1995).

58. See Blum et al. (1990). "Race" was assumed to be biologically unproblematic in the sense that it was assumed that no significant interbreeding had taken place between the so-called races.

59. See Pato et al. (1993) for the meta-analysis. The meta-analysis concludes that evidence suggesting a link between alcoholism and the dopamine D2 receptor is slightly better than the evidence against it.

60. See Lander and Schork (1994).

61. This claim should not be conflated with one that there is a possibility for genetic intervention (for any manipulative purpose, such as curing a disease) in such cases. The multiplicity of loci makes any such procedure at least technologically unfeasible – see Chapter 7.

CHAPTER 6

1. See, e.g., J. S. Haldane (1906, 1914), Loeb (1912), or Hogben (1930).

2. For historical detail on J. S. Haldane, see Douglas (1963) and Sturdy (1988); on Hopkins, see Needham and Baldwin (1949).

3. See, e.g., Baldwin (1947).

4. These included the extensive research programs of Pauling at the California Institute of Technology and of Perutz at the Medical Research Council Laboratory at Cambridge. Both were central to the development of molecular biology. The only scholarly general history of the development of molecular biology remains Olby (1974), and that work stops in the mid-1950s. A journalistic – but,

nevertheless, very useful – account that takes the history into the mid-1960s is Judson (1979). There has been little historical or philosophical attention to molecular biology since 1970.

5. See Pauling and Corey (1950). For historical detail, see Olby (1974) and Judson (1979).

6. For a history of the development of classical (that is, premolecular) genetics, see Carlson (1966). Olby (1974) and Judson (1979) cover the later period.

7. For the relevant work of Franklin and Wilkins, see Franklin and Gosling (1953) and Wilkins, Stokes, and Wilson (1953).

8. Sarkar (1996) provides a detailed account of the introduction of the concept of information into molecular biology, which only began in 1953.

9. The diversity of the origins of molecular biology has been emphasized, in particular, by Burian (1996). See, also, Zallen (1996).

10. See, especially, Thaler (1996).

11. See Sarkar (1996) for a detailed analysis of this point, and those that follow in this paragraph. For a different but complementary account of these developments, see Keller (1995). Judson (1979) has also analyzed the role played by information in molecular biology in the 1950s.

12. See Jacob et al. (1960) and Jacob and Monod (1961).

13. This will be discussed in further detail in § 6.2. There is no evidence that cybernetic reasoning played any role in the original construction of the operon model.

14. The best-known such attempt was that of Britten and Davidson (1969). Its failure has been analyzed in Sarkar (1996).

15. In fact, the only generally unsuccessful program in molecular biology has been one that was explicitly antireductionist. This is the attempt by Delbrück (e.g., 1949) to follow Bohr's (1933) suggestion and find "complementarity" in biology. The philosophical significance of this program has been systematically analyzed by Stein (1958), its failure noted by Sarkar (1989). The significance of this program in the history of molecular biology remains a matter of controversy (see, e.g., Kendrew (1967) and Stent (1968)).

16. See Pauling and Corey (1951).

17. See Monod, Changeoux, and Jacob (1963) and Monod, Wyman, and Changeux (1965). Monod (1971) even attempts, not very plausibly, a cybernetic interpretation of allostery.

18. The point that molecular biology does not have a unifying theory has often been made in the past [e.g., Hull (1972, 1974, 1976), Wimsatt (1976b), Sarkar (1989)]; that it has fragmentary theories has been noted by Sarkar (1992) and Schaffner (1993a,b); and that it is an intensely theoretical science has been emphasized by Burian (1996).

19. For an attempt in that direction, see Sarkar (1996).

20. This historical assessment is stated more strongly, for the sake of narrative simplicity, than can be strictly defended. As stated, it suggests that the molecularization of biology is to be entirely accounted for by epistemic, rather than sociological, sociopolitical, or other factors. This is a stronger thesis than is

warranted by the various narratives of the development of molecular biology that have recently been constructed [see, e.g., Keller (1995)]. However, what does appear to be strictly defensible, even given the insights of these narratives, is that epistemic factors provide a major reason for the molecularization of biology. Moreover, what does not appear to be at all defensible are those accounts that tend to ignore epistemic factors almost altogether.

21. This point has been forcefully argued by Gilbert in several works [see, e.g., Gilbert (1996)].

22. For more details on these examples, see Owen and Lamb (1990, pp. 32–3). The data reported is from crystallographic studies. Up till now, about ten antigen-Fab fragments (antigen-binding fragments of antibodies) have been crystallized and the crystal structure solved. All show the same pattern of interaction, that is, the lock-and-key fit.

23. For more detail on this example, see Tanford (1980).

24. See, for example, Cantor and Schimmel (1980a), p. 287.

25. See Tanford (1980, p. 2).

26. See Cleveland et al. (1983) for details of the colicin example.

27. Alternative models were presented by Koshland, Némethy, and Filmer (1966). For a recent review, see Perutz (1990).

28. See, e.g., J. S. Haldane (1906, 1914).

29. See Pauling et al. (1949).

30. For more details, see Dickerson and Geis (1983), pp. 125–45 and Edelstein (1986).

31. For the fourteen-strand model, see Dykes, Crepeau, and Edelstein (1978); Crepeau et al. (1978), and Edelstein (1980). For the sixteen-strand model, see Wellems and Josephs (1980); Wellems, Vassar, and Josephs (1981); and Vassar, Potel, and Josephs (1982).

32. This would be the philosophical interpretation of the claims associated with the Hopkins school (see § 6.1) and its intellectual descendants.

33. Similarly, this would be the philosophical interpretation of the structural explanations originating with Pauling and his collaborators (§ 6.1). It is also the view endorsed by Schrödinger (1944) and Weyl (1949).

34. Steric hindrance must be avoided even within the macromolecules.

35. See, e.g., Dalton ([1808] 1964), p. 164.

36. The electron-pair corresponds to a concentration of the Laplacian of the charge density field, rather than the field itself [see Bader (1990), pp. 248–64]. Additionally, even atoms (and certainly groups of atoms) form interacting systems. Quantum-mechanical states representing them are "entangled" [Shimony (1987); Sarkar (1989, 1992)]. As both Shimony and Sarkar have emphasized, this by itself poses another independent problem for reduction.

37. Cartwright (1983) has discussed other similar problems with the relation of quantum mechanics to the behavior of everyday objects (including measuring devices).

38. The value for the "hydrophobic bond" is from Stryer (1988, p. 8). All other values are from Alberts et al. (1994), p. 91.

39. "Structure determines function" is a standard locution in molecular biology and is stated as such for that reason. "Function," here, is to be construed as all behavior and not only as those behaviors that can justifiably be called "functions" because they enhance fitness or some other such reason [see, e.g., Wimsatt (1972)]. Sarkar (1991) applies Wimsatt's account to functional explanation in molecular biology.

40. See Colman (1988) and Davies, Sheriff, and Padlan (1988).

41. See, e.g., Schaffner (1967a,b, 1969, 1974, 1976, 1993a,b), Ruse (1971, 1973, 1976), Hull (1972, 1974, 1976, 1981), Wimsatt (1976b), Darden and Maull (1977), Maull (1977), Rosenberg (1978, 1985, 1994), Kitcher (1984), and Waters (1990); for a contrast, see Sarkar (1989, 1991, 1992), who points out the importance of reduction in areas of molecular biology other than genetics, for instance, cell biology.

42. Hull (1972, 1974) went even further and claimed that molecular genetics was replacing classical genetics. A similar claim was made, earlier, by Ruse (1971). However, Ruse (1976) argued in favor of reduction, as noted later in the text.

43. However, as Schaffner (1974, 1993a) has empasized, his model is supposed to capture the structure of the explanations that emerge from molecular genetics, but the satisfaction of the model in not actively pursued in research – he calls this the "peripherality thesis." This book ignores questions about the nature of research strategies. However, Sarkar (1989, 1992) has argued that Schaffner's peripherality thesis holds at best for his model of reduction, not for models based on nonformal criteria such as those emphasized in this book, but which either reject a necessary role for theories (in reduction) or are not formalized according to logical empiricist strictures about the form of theories.

44. For a discussion of the various models, see Chapter 2.

45. The classification here corresponds only partly to that of Waters (1990). He explicitly mentions (ii), which he calls "splintering." (iii) is something like his "gory details" objection. He subsumes both of these under the rubric of "explanatory incompleteness." (v) corresponds, somewhat, to his "unconnectability" objection; at least the substantive concerns are the same.

46. Kitcher (1984) does not even find theories in classical genetics.

47. This point has been particularly emphasized by Waters (1990).

48. For the best-known example of treating these extraneous issues ("extraneous" from the point of view developed in Chapter 3 of this book) as if they were central to the problem of reduction, see Kitcher (1984).

49. However, since Hull (1972, 1974) denied that molecular biology had well-developed or articulated theories, the reference to "theories" in this argument should not be taken literally. It would suffice, but be awkward, to say that classical genetics provides mechanisms for transmission whereas molecular genetics provides mechanisms for both transmission and expression.

50. Both of these points have been previously emphasized by Wimsatt (1976b), Sarkar (1989, 1992), and Waters (1990) among others.

51. Kincaid considers the reduction of molecular biology to biochemistry and, for this and other reasons, finds it wanting. However, Kincaid's characterization of molecular biology is in informational terms ("information," "signals," etc.), which is itself suspect (Sarkar 1996). Moreover, as pointed out at the beginning of this chapter, biochemistry is not the appropriate **F**-realm to which putative reductions are taking place.

52. See, for instance, Griffiths et al. (1993, p. 567, emphasis added): "The application of the Holliday model, or the Meselson-Radding model [both of which are models for recombination that will be discussed in § 6.6] ... nicely *explain* several genetic results." Claims of this sort are ubiquitous in the literature of molecular biology.

53. Recall the discussion in § 6.1.

54. There is an independent problem, that of the definition of a "classical gene." It is generally conceded that the units picked out by various claims about the action of genes (in classical genetics) and those picked out by the transmission rules are not strictly the same. This, in turn, confounds attempts at a definition in molecular terms. For particularly valuable discussions of this and related issues, see Falk (1986) and Fogle (1990). This problem will not be considered here since the importance of definitions for reduction is itself being downgraded in this book.

55. See Berg and Howe (1989) for details.

56. For more detail, see Carlson (1991) and Portin (1993).

57. See Smith, Patton, and Nadal-Ginard (1989) and Hodge and Bernstein (1994) for reviews and more detailed discussions.

58. These are also present in prokaryotes. They were fist discovered in the bacteriophage ϕX174 [see Portin (1993)].

59. See Cattaneo (1991) for a review.

60. See Fox (1987) for a review.

61. This is a *scheme* in the sense that the individual nodes would have to be specified and details filled in to generate actual explanations.

62. For a recent accessible textbook treatment that provides much greater detail, see Alberts et al. (1994).

63. This is a critical point that antireductionists such as Kitcher (1984) and Rosenberg (1978, 1985, 1994) ignore when they suggest that the elucidation of molecular mechanisms is largely a matter of redescription rather than explanation.

64. What counts as explanation depends on context – recall the discussion of Chapter 1, § 1.3. No doubt, during the 1900–30 period, the cytological account served a critical explanatory purpose (within classical genetics).

65. For more details, see Kornberg and Baker (1992). The account given here draws heavily from Chapter 15 of that work.

66. The treatment here follows Griffiths et al. (1993), Chapter 19. For the original work, see Holliday (1964, 1990) and Meselson and Radding (1975). For more detail see, for example, Low (1988).

67. For a particularly clear summary, see Griffiths et al. (1993), Chapters 7–9. The modulation of mutagenesis has been one of the most intriguing recent discoveries of biology.

68. For instance, Rosenberg (1985, p. 101) worries that the "pathway to red eye pigment production begins at many distinct molecular genes and proceeds through several alternative branched pathways. Some of the genes from which it begins are redundant, in that even if they are prevented from functioning the pigment will be produced. Others are interdependent, so that if one is blocked the other will not produce any product. Still others are "ambiguous" – belonging to several distinct pathways to different phenotypes." The reference to "molecular" genes in this passage is gratuitous. What Rosenberg is noting – in this order – are the standard phenomena about variable expressivity, penetrance, epistasis, and pleiotropy, all relevant to the pathway from **1** to **3** and irrelevant to **b**.

69. This is discussed by Shimony (1987, 1989) and Sarkar (1989, 1992).

70. However, see Sarkar (1988, 1991) for arguments designed to show that functional explanations in molecular biology may in many cases be recast as reductionist explanations.

71. A particularly clear recent exposition is that of Richards (1991).

72. See Cantor and Schimmel (1980a), p. 66.

73. The basic idea goes back to Levinthal (1968).

74. See Landry and Gierasch (1994) and Lund (1994).

75. If all chaperones are also proteins (as is currently believed), one could still choose to maintain a modified sequence hypothesis that states that all the sequences in the cell together determine (or explain) the conformation of any protein. This does not have quite the majesty of Crick's original hypothesis.

76. Whether a failure of the sequence hypothesis should be regarded as a demonstration of the insolubility of the protein-folding problem is largely a matter of taste. It depends on what one takes the folding problem to be, simply that of explaining (or predicting) protein conformations from physical principles alone, or that of doing so from sequence information and physical interactions alone. There is no consensus on this point.

77. See Correns (1900).

78. See Hull (1976).

79. See, also, Fisher (1930, 1934b). Provine (1992) gives details of the dispute between Fisher and Wright.

80. See, e.g., Crick (1958).

81. Moreover, even if Schaffner's attempt to characterize dominance at the level of the amino acid residue sequences is sometimes valid, for instance, in the case of the sickle cell trait (and even that is doubtful), it does not justify the definition he states. Interpreting his "=" as "if and only if" (which is the least of the conditions it should satisfy), it shows that whereas the "if" statement is true, the "only if" statement is false. This underscores the point made in Chapter 2, § 2.5, about the

general irrelevance of the "only if" statements in the so-called "bridge laws" or "reduction functions."

82. This assumption is made because most mutations are empirically found to be recessive. Strictly speaking, it does not follow from this that all (or even most) recessive traits arose by new mutation.

83. See, however, Cornish-Bowden (1987) for an example of a linear network without dominance [and Kacser (1987) for a reply].

84. Moreover, see Savageau (1992) for a discussion of other situations that produce negative C_e^J and for cases when Equation (6.7.2) does not hold.

85. See, e.g., Jerne (1974) and Perelson (1989).

86. See, e.g., Clarke, Mittenthal, and Sens (1993). For a different, and more promising development of genetic network theory, see McAdams and Shapiro (1995).

87. See, e.g., Duvdevani-Bar and Segel (1994).

88. See, especially, Lewontin (1992).

89. See, for example, Gilbert (1992). The strategy for the creation of a "theoretical biology" that Gilbert suggests is one that combines pattern recognition with physical reduction. Nevertheless, it is assumed that the appropriate basis from which explanation should start is the genetic level, that is, DNA. No justification is offered for this assumption.

CHAPTER 7

1. For a particularly intriguing account of these developments see Sapp (1987), which argues that a "nucleocentric" orientation that ignored non-nuclear factors came to dominate biology by midcentury, much to its detriment. One consequence of this was the dominance of genetics over other areas of biology. This pattern persisted in molecular biology, as noted in Chapter 6, as molecular genetics has assumed the central role in the new discipline since the 1960s.

2. See, e.g., Cook-Deegan (1994, p. 176). This is a partisan work defending the HGP but, nevertheless, admits the detrimental effect of the HGP with respect to the funding of other types of biological research in the United States.

3. This story is much more complex than is usually acknowledged; see Paul (1995).

4. Such a position, that is, the biological basis of (human) sexual orientation was originally (controversially) urged by LeVay (1993) in the case of male homosexuality without any explicit claim of a genetic etiology. Subsequently, after the report of Hamer et al. (1993), LeVay accepted the genetic etiology (see LeVay and Hamer (1994)). For important criticism of the entire project, see Byne (1994).

5. See, e.g., Freeman and Lundelius (1982).

6. A full appreciation of this point, which goes back at least to Hogben (1933), is exactly what continues to be missing in modern attempts to distinguish between the effects of genes and the environment even when there is explicit acknowledgement that genotype–environment interactions are important. See, for example,

Bouchard (1994) and Plomin (1994), both of whom explicitly acknowledge these interactions but nevertheless attempt to ascribe precise numerical fragments of influence to the genotype and the environment.

7. This is the case for the so-called continuous or quantitative traits that are believed to be produced by a very large number of alleles each with a very small influence.

8. Note, however, that an "explication" does not necessarily require an explicit definition. Implicit definitions of terms by their incorporation into a set of precise rules that govern the use of those terms are sufficient.

9. See, for example, Smith (1992).

10. See Carnap (1950) for the account of explication on which this discussion is based.

11. Contemporary reductionists of this sort include Schaffner (1993a,b) and Waters (1990); anti-reductionists include Mayr (1982).

12. See, especially, Causey (1977), Hooker (1981a,b,c), and Schaffner (1993a,b).

References

Adams, E. W. 1959. "The Foundations of Rigid Body Mechanics and the Derivation of Its Laws from Particle Mechanics." In Henkin, L., Suppes, P., and Tarski, A. (eds.), *The Axiomatic Method*. Amsterdam: North-Holland, pp. 250–65.

Alberts, B., Bray, D., Lewis, J., Raff, M., Roberts, K., and Watson, J. D. 1994. *Molecular Biology of the Cell*. 3rd. ed., New York: Garland.

Allen, G. 1979. *Thomas Hunt Morgan. The Man and His Science*. Princeton, NJ: Princeton University Press.

Armstrong, D. M. 1983. *What Is a Law of Nature?* Cambridge, UK: Cambridge University Press.

Atkins, J. F., Weiss, R. B., Thompson, S., and Gesteland, R. F. 1991. "Towards a Genetic Dissection of the Basis of Triplet Decoding, and Its Natural Subversion: Programmed Reading Frame Shifts and Hops." *Annual Review of Genetics* **25**: 201–28.

Avery, O. T., Macleod, C. M., and McCarty, M. 1944. "Studies on the Chemical Nature of the Substance Inducing Transformation of Pneumococcal Types: Induction of Transformation by a Deoxyribonucleic Acid Fraction Isolated from Pneumococcus III." *Journal of Experimental Medicine* **79**:137–57.

Bader, R. F. W. 1990. *Atoms in Molecules: A Quantum Theory*. Oxford: Clarendon Press.

Bailey, J. M. 1995. "Sexual Orientation Revolution." *Nature Genetics* **11**:353–4.

Baldwin, E. 1947. *Dynamic Aspects of Biochemistry*. Cambridge, UK: Cambridge University Press.

Balzer, W., and Dawe, C. M. 1986a. "Structure and Comparison of Genetic Theories: (1) Classical Genetics." *British Journal for the Philosophy of Science* **37**:55–69.

Balzer, W., and Dawe, C. M. 1986b. "Structure and Comparison of Genetic Theories: (2) The Reduction of Character-Factor Genetics to Molecular Genetics." *British Journal for the Philosophy of Science* **37**:177–191.

Balzer, W., and Sneed, J. D. 1977. "Generalized Net Structures of Empirical Theories I." *Studia Logica* **36**:195–211.

Baron, M., Endicott, J., and Ott, J. 1990. "Genetic Linkage in Mental Illness." *British Journal of Psychiatry* **157**:645–55.

References

Baron, M., Freimer, N. F., Risch, N., Lerer, B., Alexander, J. R., Straub, R. E., Asokan, A., Das, K., Peterson, A., Amos, A., Endicott, J., Ott, J., and Gilliam, T. C. 1993. "Diminished Support for Linkage Between Manic Depressive Illness and X-Chromosome Markers in Three Israeli Pedigrees." *Nature Genetics* **3**: 49–55.

Barr, W. F. 1971. "A Syntactic and Semantic Analysis of Idealizations in Science." *Philosophy of Science* **38**:258–72.

Barr, W. F. 1974. "A Pragmatic Analysis of Idealizations in Physics." *Philosophy of Science* **41**:48–64.

Bateson, W. 1906. "An Address on Mendelian Heredity and Its Application to Man." *Brain* **29**:157–79.

Beadle, G. W., and Tatum, E. 1941. "Genetics and Metabolism in Neurospora." *Proceedings of the National Academy of Sciences (USA)* **27**:499–506.

Bell, A. E. 1977. "Heritability in Retrospect." *Journal of Heredity* **68**:297–300.

Bell, J., and Haldane, J. B. S. 1937. "The Linkage between the Genes for Colour-blindness and Haemophilia in Man." *Proceedings of the Royal Society B* **123**:119–50.

Berg, D. E., and Howe, M. M. (eds.) 1989. *Mobile DNA*. Washington: American Society for Microbiology.

Bernal, J. D. 1939. "Structure of Proteins." *Proceedings of the Royal Institution* **30**:541–57.

Block, N. J., and Dworkin, G. (eds.) 1976a. *The IQ Controversy: Critical Readings*. New York: Pantheon.

Block, N. J., and Dworkin, G. 1976b. "IQ, Heritability, and Inequality." In Block, N. J., and Dworkin, G. (eds.), New York: Pantheon. pp. 410–542.

Blum, K., Noble, E. P., Sheridan, P. J., Montgomery, A., Ritchie, T., Jagadeeswaran, P., Nogami, H., Briggs, A. H., and Cohm, J. B. 1990. "Allelic Association of Human Dopamine D_2 Receptor Gene in Alcoholism." *Journal of the American Medical Association* **263**:2055–60.

Bock, G. R., and Goode, J. A. (eds.) 1996. *Genetics of Criminal and Antisocial Behavior*. Chicester, UK: John Wiley & Sons.

Bohr, N. 1933. "Light and Life." *Nature* **131**:421–3, 457–9.

Bonevac, D. A. 1982. *Reduction in the Abstract Sciences*. Indianapolis, IN: Hackett.

Bookstein, F. L. 1990. "An Interaction Effect is Not a Measurement." *Behavioral and Brain Sciences* **13**:121–2.

Botstein, D., White, R. L., Skolnick, M. H., and Davies, R. W. 1980. "Construction of a Genetic Linkage Map in Man Using Restriction Fragment Length Polymorphisms." *American Journal of Human Genetics* **32**:314–31.

Bouchard, T. J. 1994. "Genes, Environment, and Personality." *Science* **264**:1700–1.

Bouchard, T. J., Lykken, D. T., McGue, M., Segal, N. L., and Tellegen, A. 1990. "Sources of Human Psychological Differences: The Minnesota Study of Twins Reared Apart." *Science* **250**:223–8.

Bourbaki, N. 1968. *Theory of Sets*. Paris: Hermann.

References

Boveri, T. 1902. "Über den Einfluss der Samenzelle auf die Larvencharaktere der Echiniden." *Verhandlungen der physikalisch-medinischen Gesselschaft zu Würzburg. Neue Folge* **35**:67–90.

Britten, R. J., and Davidson, E. H. 1969. "Gene Regulation for Higher Cells: A Theory." *Science* **165**:349–57.

Burian, R. M. 1996. "Underappreciated Pathways Toward Molecular Genetics as Illustrated by Jean Brachet's Cytochemical Embryology." In Sarkar, S. (ed.) *The Philosophy and History of Molecular Biology: New Perspectives*. Dordrecht: Kluwer. pp. 67–85.

Byne, W. 1994. "The Biological Evidence Challenged." *Scientific American* **270**:50–5.

Cantor, C. R., and Schimmel, P. R. 1980a. *Biophysical Chemistry. Part I: The Conformation of Biological Macromolecules*. San Francisco: W. H. Freeman.

Cantor, C. R., and Schimmel, P. R. 1980b. *Biophysical Chemistry. Part III: The Behavior of Biological Macromolecules*. San Francisco: W. H. Freeman.

Carlson, E. A. 1966. *The Gene: A Critical History*. Philadelphia: W. B. Saunders.

Carlson, E. A. 1991. "Defining the Gene: An Evolving Context." *American Journal of Human Genetics* **49**:475–87.

Carnap, R. 1934. *The Unity of Science*. London: Kegan Paul, Trench, Trubner & Co.

Carnap, R. 1937. *The Logical Syntax of Language*. London: Kegan Paul, Trench, Trubner & Co.

Carnap, R. 1942. *Introduction to Semantics*. Cambridge, MA: Harvard University Press.

Carnap, R. 1950. "Empiricism, Semantics, and Ontology." *Revue internationale de philosophie* **4**:20–40.

Carnap, R. 1956. "The Methodological Character of Theoretical Concepts." In Feigl, H., and Scriven, M. (eds.), *The Foundations of Science and the Concepts of Psychology and Psychoanalysis*. Minneapolis, MN: University of Minnesota Press. pp. 38–76.

Carnap, R. 1963. "Replies and Systematic Expositions." In Schilpp, P. A. (ed.), *The Philosophy of Rudolf Carnap*. La Salle: Open Court, IL. pp. 859–1013.

Carnap, R. 1977. *Two Essays on Entropy*. Berkeley: University of California Press.

Cartwright, N. 1983. *How the Laws of Physics Lie*. New York: Oxford University Press.

Cartwright, N. 1989. *Nature's Capacities and Their Measurement*. New York: Oxford University Press.

Castle, W. E. 1903. "Mendel's Law of Heredity." *Science* **18**:396–406.

Cattaneo, R. 1991. "Different Types of Messenger RNA Editing." *Annual Review of Genetics* **25**:71–88.

Causey, R. L. 1972a. "Attribute-Identities in Microreduction." *Journal of Philosophy* **69**:407–22.

Causey, R. L. 1972b. "Uniform Microreductions." *Synthese* **25**:176–218.

Causey, R. L. 1977. *The Unity of Science*. Dordrecht: Reidel.

References

Church, A. 1956. *Introduction to Mathematical Logic*. Princetion, NJ: Princeton University Press.

Churchland, P. 1979. *Scientific Realism and the Plasticity of Mind*. Cambridge, UK: Cambridge University Press.

Churchland, P. 1981. "Eliminative Materialism and the Propositional Attitudes." *The Journal of Philosophy* **78**:67–90.

Churchland, P. 1984. *Matter and Consciousness: A Contemporary Introduction to the Philosophy of Mind*. Cambridge, MA: MIT Press.

Churchland, P. S. 1986. *Neurophilosophy*. Cambridge, MA: MIT Press.

Clarke, B., Mittenthal, J. E., and Senn, M. 1993. "A Model for the Evolution of Networks of Genes." *Journal of Theoretical Biology* **165**:269–89.

Cleveland, M. v., Slatin, S., Finkelstein, A., and Levinthal, C. 1983. "Structure-Function Relationships for a Voltage-Dependent Ion Channel: Properties of COOH-Terminal Fragments of Colicin E1." *Proceedings of the National Academy of Science USA* **80**:3706–10.

Cloninger, C. R. 1987. "Neurogenetic Adaptive Mechanisms in Alcoholism." *Science* **236**:410–16.

Colman, P. M. 1988. "Structure of Antibody-Antigen Complexes: Implications for Immune Recognition." *Advances in Immunology* **43**:99–132.

Cook-Deegan, R. 1994. *The Gene Wars*. New York: Norton.

Cornish-Bowden, A. 1987. "Dominance is not Inevitable." *Journal of Theoretical Biology* **125**:333–8.

Correns, C. 1900. "G. Mendels Regel über das Verhalten der Nachkommenshaft der Rassenbastarde." *Berichte der deutschen botanischen Gesellschaft* **18**:158–68.

Crepeau, R. H., Dykes, G., Garrell, R., and Edelstein, S. J. 1978. "Diameter of Haemoglobin S Fibres in Sickled Cells." *Nature* **274**:616–17.

Crick, F. H. C. 1958. "On Protein Synthesis." *Symposium of the Society for Experimental Biology* **12**:138–63.

Crow, J. F. 1990. "How Important Is Detecting Interaction?" *Behavioral and Brain Sciences* **13**:126–7.

Crusio, W. E. 1990. "Estimating Heritabilities in Quantitative Behavior Genetics: A Station Passed." *Behavioral and Brain Sciences* **13**:127–8.

Dalton, J. [1808] 1964. *A New System of Chemical Philosophy*. New York: Citadel Press.

Darden, L. 1991. *Theory Change in Science: Strategies from Mendelian Genetics*. New York: Oxford University Press.

Darden, L., and Maull, N. 1977. "Interfield Theories." *Philosophy of Science* **44**:43–64.

Darnell, J., Lodish, H., and Baltimore, D. 1990. *Molecular Cell Biology*. 2nd. ed. New York: Scientific American Books.

Davenport, C. 1915. *The Feebly Inhibited; Nomadism or the Wandering Impulse, with Special Reference to Heredity, Inheritance of Temperament*. Washington: Carnegie Institution.

Davidson, D. 1970. "Mental Events." In Foster, L., and Swanson, J. W. (eds.), *Experience and Theory*. Amherst: University of Massachusetts Press. pp. 79–101.

References

Davies, D. R., Sheriff, S., and Padlan, E. A. 1988. "Antibody-Antigen Complexes." *Journal of Biological Chemistry* **263**:10541–4.

Delbrück, M. 1949. "A Physicist Looks at Biology." *Transactions of the Connecticut Academy of Sciences* **38**:173–90.

Devlin, B., Fienberg, S. E., Resnick, D. P., and Roeder, K. (eds.) 1997. *Intelligence, Genes, and Success.* New York: Springer-Verlag.

Dickerson, R. E., and Geis, I. 1983. *Hemoglobin: Structure, Function, Evolution, and Pathology.* Menlo Park, CA: Benjamin/Cummings.

Diggs, L. W., and Ching, R. E. 1934. "Pathology of Sickle Cell Anemia." *Southern Medical Journal* **27**:839–44.

Dillon, L. S. 1987. *The Gene: Its Structure, Function, and Evolution.* New York: Plenum.

Dobzhansky, T., and Spassky, B. 1944. "Genetics of Natural Populations: XI. Manifestation of genetic variants in *Drosophila pseudoobscura* in different environments." *Genetics* **29**:270–90.

Douglas, C. G. 1963. "John Scott Haldane, 1860–1936." In Cunningham, D. J. C., and Lloyd, B. B. (eds.), *The Regulation of Human Respiration.* Oxford: Blackwell, pp. 4–32.

Dudley, R. M. 1990. "Interaction and Dependence Prevent Estimation." *Behavioral and Brain Sciences* **13**:132–3.

Duvdevani-Bar, S., and Segel, L. 1994. "On Topological Simulations in Developmental Biology." *Journal of Theoretical Biology* **166**:33–50.

Dykes, G., Crepeau, R. H., and Edelstein, S. J. 1978. "Three-dimensional Reconstruction of the Fibres of Sickle Cell Hemoglobin." *Nature* **272**:506–10.

Earman, J. 1986. *A Primer on Determinism.* Dordrecht: Reidel.

Edelstein, S. J. 1980. "Patterns in the Quinary Structures of Proteins. Plasticity and Inequivalence of Individual Molecules in Helical Arrays of Sickle Cell Hemoglobin and Tubulin." *Biophysical Journal* **32**:347–57.

Edelstein, S. J. 1986. *The Sickled Cell: From Myths to Molecules.* Cambridge, MA: Harvard University Press.

Editorial. 1995. "Crimes against Genetics." *Nature Genetics* **11**:223–4.

Egeland, J. E., Gerhard, D. S., Pauls, D. L., Sussex, J. N., Kidd, K. K., Allen, C. R., Hostetter, A. M., and Housman, D. E. 1987. "Bipolar Affective Disorders Linked to DNA Markers Linked on Chromosome 11." *Nature* **325**:783–87.

Einstein, A. 1905.

Enç, B. 1976. "Identity Statements and Microreductions." *Journal of Philosophy* **73**:285–306.

Falconer, D. S. 1989. *Introduction to Quantitative Genetics.* 3rd. ed. New York: Open Court.

Falk, R. 1986. "What is a Gene?" *Studies in the History and Philosophy of Science* **17**:133–73.

Falk, R., and Sarkar, S. 1991. "The Real Objective of Mendel's Paper: A Response to Monaghan and Corcos." *Biology and Philosophy* **6**:447–51.

Fancher, R. E. 1985. *The Intelligence Men: Makers of the I. Q. Controversy.* New York: Norton.

References

Feigl, H. 1963. "Physicalism, Unity of Science and the Foundations of Psychology." In Schilpp, P. A. (ed.), *The Philosophy of Rudolf Carnap*. La Salle: Open Court, IL. pp. 227–69.

Feldman, M. W., and Lewontin, R. C. 1975. "The Heritability Hang-up." *Science* **190**:1163–8.

Feyerabend, P. 1962. "Explanation, Reduction, and Empiricism." *Minnesota Studies in the Philosophy of Science* **3**:28–97.

Fine, A. 1990. "Causes of Variability: Disentangling Nature and Nurture." *Midwest Studies in Philosophy* **15**:94–113.

Fischer, E. P. and Lipson, C. 1988. *Thinking About Science: Max Delbrück and the Origins of Molecular Biology*. New York: Norton.

Fisher, R. A. 1918. "The Correlation between Relatives on the Supposition of Mendelian Inheritance." *Transactions of the Royal Society of Edinburgh* **52**:399–433.

Fisher, R. A. 1928a. "The Possible Modification of the Response of the Wild Type to Recurrent Mutations." *American Naturalist* **62**:115–26.

Fisher, R. A. 1928b. "Two Further Notes on the Origin of Dominance." *American Naturalist* **62**:571–4.

Fisher, R. A. 1930. *The Genetical Theory of Natural Selection*. Oxford: Clarendon Press.

Fisher, R. A. 1934a. "The Effect of Methods of Ascertainment upon the Estimation of Frequencies." *Annals of Eugenics* **6**:13–25.

Fisher, R. A. 1934b. "Professor Wright on the Theory of Dominance." *American Naturalist* **68**:370–4.

Fisher, R. A. 1951. "Limits to Intensive Production in Animals." *British Agricultural Bulletin* **4**:217–18.

Fodor, J. A. 1974. "Special Sciences (or: The Disunity of Science as a Working Hypothesis)." *Synthese* **28**:97–115.

Fogle, T. 1990. "Are Genes Units of Inheritance?" *Biology and Philosophy* **5**:349–71.

Fox, T. D. 1987. "Natural Variation in the Genetic Code." *Annual Reveiw of Genetics* **21**:67–91.

Franklin, R. E., and Gosling, R. G. 1953. "Molecular Configuration in Sodium Thymonucleate." *Nature* **171**:740–1.

Freeman, G., and Lundelius, J. W. 1982. "The Developmental Genetics of Dextrality and Sinistrality in the Gastropod *Lymnaea peregra*." *Wilhelm Roux's Archives* **191**:69–83.

Friedman, K. 1982. "Is Intertheoretic Reduction Feasible?" *British Journal for the Philosophy of Science* **33**:17–40.

Froggatt, P., and Nevin, N. C. 1971. "The 'Law of Ancestral Heredity' and the Mendelian-Ancestrian Controversy in England, 1889–1906." *Journal of Medical Genetics* **8**:1–36.

Galton, F. 1865. "Hereditary Talent and Character." *Macmillan's Magazine* **12**:157–66; 318–27.

Galton, F. 1889. *Natural Inheritance*. London: Macmillan.

References

Garrod, A. 1902. "The Incidence of Alkaptonuria: A Study in Chemical Individuality." *Lancet* 13 December:1616–20.

Garrod, A. E. 1909. *Inborn Errors of Metabolism*. London: Frowde, Hodder, and Stroughton.

Gifford, F. 1990. "Genetic Traits." *Biology and Philosophy* **5**:327–47.

Gilbert, S. 1996. "Enzymatic Adaptation and the Entrance of Molecular Biology into Embryology." In Sarkar, S. (ed.), *The Philosophy and History of Molecular Biology: New Perspectives*. Dordrecht: Kluwer. pp. 101–23.

Gilbert, W. 1992. "A Vision of the Grail." In Kevles, D. J., and Hood, L., (ed.), *The Code of Codes*. Cambridge, MA: Harvard University Press. pp. 83–97.

Glymour, C. 1984. "Explanation and Realism." In Leplin, J. (ed.), *Scientific Realism*. Berkeley: University of California Press. pp. 173–92.

Goldschmidt, R. B. 1938. *Physiological Genetics*. New York: McGraw-Hill.

Goldschmidt, R. B. 1955. *Theoretical Genetics*. Berkeley: University of California Press.

Gould, S. J. 1996. *The Mismeasure of Man*. 2nd. ed. New York: W. W. Norton.

Hahn, E. V., and Gillespie, E. B. 1927. "Sickle Cell Anemia." *Archives of Internal Medicine* **39**:233–54.

Haldane, J. B. S. 1930. "A Note on Fisher's Theory of the Origin of Dominance, and a Correlation Between Dominance and Linkage." *American Naturalist* **64**:87–90.

Haldane, J. B. S. 1936a. "A Provisional Map of a Human Chromosome." *Nature* **137**:398–400.

Haldane, J. B. S. 1936b. "Some Principles of Causal Analysis in Genetics." *Erkenntnis* **6**:346–57.

Haldane, J. B. S. 1938a. "Blood Royal: A Study of Haemophilia in the Royal Families of Europe." *Modern Quarterly* **1**:129–39.

Haldane, J. B. S. 1938b. "The Estimation of Frequencies of Recessive Conditions in Man." *Annals of Eugenics* **8**:255–62.

Haldane, J. B. S. 1939. *The Marxist Philosophy and the Sciences*. New York: Random House.

Haldane, J. B. S. 1942. *New Paths in Genetics*. New York: Random House.

Haldane, J. S. 1906. "Life and Mechanism." *Guy's Hospital Reports* **60**:89–123.

Haldane, J. S. 1914. *Mechanism, Life, and Personality: An Examination of the Mechanistic Theory of Life and Mind*. London: J. Murray.

Hamer, D. H., Hu, S., Magnuson, V. L., Hu, N., and Pattatucci, A. M. L. 1993. "A Linkage Between DNA Markers on the X Chromosome and Male Sexual Orientation." *Science* **261**:321–27.

Hamlett, G. W. D. 1926. "The Linkage Disturbance Involved in the Chromosome Translocation I. of Drosophila, and Its Probable Significance." *Biological Bulletin of the Marine Biological Laboratory, Woods Hole* **51**:435–42.

Harding, S. 1975. "Making Sense of Observation Sentences." *Ratio* **17**:65–71.

Hempel, C. G. 1969. "Reduction: Linguistic and Ontological Issues." In Morgenbesser, S., Suppes, P., and White, M. (eds.), *Philosophy, Science and Method: Essays in Honor of Ernest Nagel*. New York: St. Martin's Press. pp. 179–99.

References

Hempel, C. G., and Oppenheim, P. 1948. "Studies in the Logic of Explanation." *Philosophy of Science* **15**:135–75.

Herrick, J. B. 1910. "Peculiar Elongated Sickle-Shaped Red Blood Corpuscles in a Case of Severe Anemia." *Archives of Internal Medicine* **6**:517–21.

Herrnstein, R. J. 1971. "IQ." *Atlantic Monthly* **228**:43–64.

Herrnstein, R. J. 1973. *IQ in the Meritocracy*. Boston: Atlantic-Little Brown.

Herrnstein, R. J., and Murray, C. 1994. *The Bell Curve: Intelligence and Class Structure in American Life*. New York: Free Press.

Hershey, A. D., and Chase, M. 1952. "Independent Functions of Viral Proteins and Nucleic Acid in Growth of Bacteriophage." *Journal of General Physiology* **36**:39–56.

Hodge, D., and Bernstein, S. I. 1994. "Genetic and Biochemical Analysis of Alternative Splicing." *Advances in Genetics* **31**:207–81.

Hogben, L. 1930. *The Nature of Living Matter*. London: Kegan Paul, Trench, Trubner.

Hogben, L. 1931. "The Genetic Analysis of Family Traits. I. Single Gene Substitutions." *Journal of Genetics* **25**:97–112.

Hogben, L. 1933. *Nature and Nurture*. New York: W. W. Norton.

Holden, C. 1995. "More on Genes and Homosexuality." *Science* **268**:1571.

Holliday, R. 1964. "A Mechanism for Gene Conversion in Fungi." *Genetical Research* **5**:282–303.

Holliday, R. 1990. "The History of the DNA Heteroduplex." *Bioessays* **12**:133–42.

Hooker, C. A. 1981a. "Towards a General Theory of Reduction. Part I: Historical and Scientific Setting." *Dialogue* **20**:38–59.

Hooker, C. A. 1981b. "Towards a General Theory of Reduction. Part II: Identity in Reduction." *Dialogue* **20**:201–36.

Hooker, C. A. 1981c. "Towards a General Theory of Reduction. Part III: Cross-Categorical Reduction." *Dialogue* **20**:496–529.

Hu, S., Pattatucci, A. M. L., Patterson, C., Li, L., Fulker, D. W., Cherny, S. S., Kruglyak, L., and Hamer, D. H. 1995. "Linkage between Sexual Orientation and Chromosome Xq28 in Males but Not in Females." *Nature Genetics* **11**:248–56.

Hull, D. 1972. "Reduction in Genetics – Biology or Philosophy?" *Philosophy of Science* **39**:491–9.

Hull, D. 1974. *Philosophy of Biological Science*. Englewood Cliffs, NJ: Prentice-Hall.

Hull, D. 1976. "Informal Aspects of Theory Reduction." *Boston Studies in the Philosophy of Science* **32**:653–70.

Hull, D. 1981. "Reduction and Genetics." *Journal of Medicine and Philosophy* **6**:125–43.

Hull, F. H. 1946. "Regression Analyses of Corn Yield Data." *Genetics* **31**:219.

Humphries, P., Kenna, P., and Farrer, G. J. 1992. "On the Molecular Genetics of Retinitis Pigmentosa." *Science* **256**:805–8.

Huxley, J. 1942. *Evolution: The Modern Synthesis*. London: George Allen & Unwin.

Ingram, V. M. 1956. "A Specific Chemical Difference between the Globins of Normal Human and Sickle-Cell Haemoglobin." *Nature* **178**:792–4.

References

Jacob, F., and Monod, J. 1961. "Genetic Regulatory Mechanisms in the Synthesis of Proteins." *Journal of Molecular Biology* **3**:318–56.

Jacob, F., Perrin, D., Sanchez, C., and Monod, J. 1960. "L'opéron: groupe de gènes à expression coodinée par un opérateur." *Comptes Rendu des Séances de l'Academie des Sciences* **250**:1727–9.

Jacquard, A. 1983. "Heritability: One Word, Three Concepts." *Biometrics* **39**:465–77.

Jensen, A. R. 1969. "How Much Can We Boost IQ and Scholastic Achievement?" *Harvard Educational Review* **39**:1–123.

Jensen, A. R. 1973. *Educability and Group Differences*. New York: Harper and Row.

Jerne, N. K. 1974. "Towards a Network Theory of the Immune System." *Ann. Immunol. (Inst. Pasteur)* **125**:373–89.

Johannsen, W. 1909. *Elemente der exacten Erblichkeitslehre.* Jena, Germany: Fischer.

Judson, H. F. 1979. *The Eighth Day of Creation*. New York: Simon and Schuster.

Kacser, H. 1987. "Dominance Not Inevitable But Very Likely." *Journal of Theoretical Biology* **126**:505–6.

Kacser, H., and Burns, J. A. 1973. "The Control of Flux." *Symposia of the Society for Experimental Biology* **27**:65–104.

Kacser, H., and Burns, J. A. 1981. "The Molecular Basis of Dominance." *Genetics* **97**:639–66.

Kamin, L. J. 1974. *The Science and Politics of I.Q.* Potomac, MD: Erlbaum Associates.

Kauffman, S. A. 1972. "Articulation of Parts Explanation in Biology and the Rational Search for Them." *Boston Studies in the Philosophy of Science* **8**:257–72.

Kay, L. 1993. *The Molecular Vision of Life: Caltech, the Rockefeller Foundation, and the Rise of the New Biology.* New York: Oxford University Press.

Keller, E. F. 1995. *Refiguring Life: Metaphors of Twentieth Century Biology.* New York: Columbia University Press.

Kelsoe, J. R., Ginns, E. I., Egeland, J. A., Gerhard, D. S., Goldstein, A. M., Bale, S. J., Pauls, D. L., Long, R. T., Kidd, K. K., Conte, G., Housman, D. E., and Paul, S. M. 1989. "Re-evaluation of the Linkage Relationship Between Chromosome 11p Loci and the Gene for Bipolar Affective Disorder in the Old Order Amish." *Nature* **342**:238–43.

Kemeny, J., and Oppenheim, P. 1956. "On Reduction." *Philosophical Studies* **7**:6–19.

Kempthorne, O. 1969. *An Introduction to Genetic Statistics*. Ames: Iowa State University Press.

Kempthorne, O. 1978. "Logical, Epistemological and Statistical Aspects of Nature-Nurture Data Interpretation." *Biometrics* **34**:1–23.

Kendrew, J. C. 1967. "How Molecular Biology Was Started." *Scientific American* **216**(3):141–4.

Kennedy, J. L., Giuffra, L. A., Moises, H. W., Cavalli-Sforza, L. L., Pakstis, A. J., Kidd, J. R., Castiglione, C. M., Sjogren, B., Wetterberg, L., and Kidd, K. K. 1988. "Evidence Against Linkage of Schizophrenia to Markers on Chromosome 5 in a Northern Swedish Pedigree." *Nature* **336**:167–9.

References

Kevles, D. J. 1986. *In the Name of Eugenics*. Berkeley: University of California Press.

Khalfa, J. (ed.) 1994. *What Is Intelligence?* Cambridge, UK: Cambridge University Press.

Khoury, M. J., Beaty, T. H., and Cohen, B. H. 1993. *Fundamentals of Genetic Epidemiology*. New York: Oxford University Press.

Kim, J. 1984. "Concepts of Supervenience." *Philosophy and Phenomenological Research* **45**:153–76.

Kim, J. 1987. "'Strong' and 'Global' Supervenience Revisited." *Philosophy and Phenomenological Research* **48**:315–26.

Kim, J. 1989. "The Myth of Nonreductive Materialism." *Proceedings and Addresses of the American Philosophical Association* **63**:31–47.

Kimbrough, S. O. 1979. "On the Reduction of Genetics to Molecular Biology." *Philosophy of Science* **46**:389–406.

Kincaid, H. 1990. "Molecular Biology and the Unity of Science." *Philosophy of Science* **57**:575–93.

Kitcher, P. 1984. "1953 and All That. A Tale of Two Sciences." *Philosophical Review* **93**:335–73.

Kitcher, P. 1989. "Explanatory Unification and the Causal Structure of the World." In Kitcher, P., and Salmon, W. C. (eds.) *Scientific Explanation*. Minneapolis: University of Minnesota Press. pp. 410–505.

Kohler, R. E. 1994. *Lords of the Fly: Drosophila Genetics and the Experimental Life*. Chicago: University of Chicago Press.

Kornberg, A., and Baker, T. 1992. *DNA Replication*. 2nd ed., New York: W. H. Freeman.

Koshland, D. E. 1990. "The Rational Approach to the Irrational." *Science* **250**:189.

Koshland, D. E., Némethy, G., and Filmer, D. 1966. "Comparison of Experimental Binding Data and Theoretical Models in Proteins Containing Subunits." *Biochemistry* **5**:365–85.

Koslowski, D. J., Bhat, G. J., Perollaz, A. L., Feagin, J. E., and Stuart, K. 1990. "The MURF3 Gene of *T. brucei* Contains Multiple Domains of Extensive Editing and is Homologous to a Subunit of NADH Dehydrogenase." *Cell* **62**:901–11.

Krafka, J. 1920. "The Effect of Temperature upon Facet Number in the Bar-eyed Mutant of Drosophila." *Journal of General Physiology* **2**:409–64.

Kuhn, T. 1962. *The Structure of Scientific Revolutions*. Chicago: University of Chicago Press.

Lander, E. S., and Botstein, D. 1989. "Mapping Mendelian Factors Underlying Quantitative Traits Using RFLP Linkage Maps. *Genetics* **121**:185–99.

Lander, E. S., and Schork, N. J. 1994. "Genetic Dissection of Complex Traits." *Science* **265**:2037–48.

Landry, S. J., and Gierasch, L. M. 1994. "Polypeptide Interactions with Molecular Chaperones and Their Relationship to In Vivo Protein Folding." *Annual Review of Biophysics and Biomolecular Structure* **23**:645–69.

Laymon, R. 1991. "Computer Simulations, Idealizations and Approximations." In Fine, A., Forbes, M., and Wessels, L., (eds.), *PSA 1990*. Vol. 2. East Lansing, MI: Philosophy of Science Association. pp. 519–34.

Layzer, D. 1974. "Heritability Analyses of IQ Scores: Science or Numerology?" *Science* **183**:1259–66.

Lederberg, J., Lederberg, E. M., Zinder, N. D., and Lively, E. R. 1951. "Recombination Analysis of Bacterial Heredity." *Cold Spring Harbor Symposia on Quantitative Biology* **16**:413–43.

Leggett, A. J. 1987. *The Problems of Physics*. Oxford: Oxford University Press.

LeVay, S. 1993. *The Sexual Brain*. Cambridge, MA: MIT Press.

LeVay, S., and Hamer, D. H. 1994. "Evidence for a Biological Influence in Male Homosexuality." *Scientific American* **270**(5):20–5.

Levinthal, C. 1966. "Molecular Model Building by Computer." *Scientific American* **214**(6):42–53.

Levinthal, C. 1968. "Are There Pathways in Protein Folding?" *Journal de Chemie Physique et de Physico-chimie Biologique* **65**:44–5.

Lewontin, R. C. 1970. "Race and Intelligence." *Bulletin of the Atomic Scientists* **26**:2–8.

Lewontin, R. C. 1974. "The Analysis of Variance and the Analysis of Causes." *American Journal of Human Genetics* **26**:400–11.

Lewontin, R. C. 1992. *Biology as Ideology: The Doctrine of DNA*. New York: Harper-Perennial.

Lloyd, E. 1988. *The Structure and Confirmation of Evolutionary Theory*. Westport, CT: Greenwood Press.

Loeb, J. 1912. *The Mechanistic Conception of Life*. Chicago: University of Chicago Press.

Loomis, W. F., and Sternberg, P. W. 1995. "Genetic Networks." *Science* **269**:649.

Low, K. B. (ed.) 1988. *The Recombination of the Genetic Material*. San Diego: Academic Press.

Lund, P. 1994. "The Chaperonin Cycle and Protein Folding." *Bioessays* **16**:229–31.

Lush, J. L. 1943. *Animal Breeding Plans*. 2nd. ed. Ames, IA: Collegiate Press.

Mann, C. C. 1994. "Behavioral Genetics in Transition." *Science* **264**:1686–9.

Marshall, E. 1995. "NIH's 'Gay Gene' Study Questioned." *Science* **268**:1841.

Martin, M. 1972. "The Body-Mind Problem and Neurophysiological Reduction." *Theoria* **37**:1–14.

Mason, V. R. 1922. "Sickle Cell Anemia." *Journal of the American Medical Association* **79**:1318–20.

Maull, N. 1977. "Unifying Science Without Reduction." *Studies in the History and Philosophy of Science* **8**:143–71.

Mayr, E. 1982. *The Growth of Biological Thought*. Cambridge, MA: Harvard University Press.

Mayr, E., and Provine, W. B. (eds.) 1980. *The Evolutionary Synthesis*. Cambridge, MA: Harvard University Press.

References

Mazumdar, P. M. H. 1992. *Eugenics, Human Genetics, and Human Failings: The Eugenics Society, Its Sources and Its Critics in Britain.* London: Routledge.

McAdams, H. H., and Shapiro, L. 1995. "Circuit Simulation of Genetic Networks." *Science* **269**:650–6.

Mendel, G. 1866. "Versuche über Pflanzenhybriden." *Verhandlungen des naturforschenden Vereines in Brunn* **4**:3–44.

Meselson, M. S., and Radding, C. M. 1975. "A General Model for Genetic Recombination." *Proceedings of the National Academy of Sciences (USA)* **72**:358–61.

Meselson, M. S., and Stahl, F. W. 1958. "The Replication of DNA in *Escherichia coli.*" *Proceedings of the National Academy of Sciences (USA)* **44**:671–82.

Moises, H. W., Yang, L., Kristbjarnarson, H., Wiese, C., Byerley, W., Macciardi, F., Arolt, V., Blackwood, D., Liu, X., Sjörgen, B., Ascháuer, H. N., Hwu, H.-G., Jang, K., Livesley, W. J., Kennedy, J. L., Zoega, T., Ivarsson, O., Bui, M.-T., Yu, M.-H., Havsteen, B., Commenges, D., Weissenbach, J., Schwinger, E., Gottesman, I. I., Pakstis, A. J., Wetterberg, L., Kidd, K. K., and Helgason, T. 1995. "An International Two-Stage Genome-Wide Search for Schizophrenia Susceptibility Genes." *Nature Genetics* **11**:321–4.

Monaghan, F. V., and Corcos, A. F. 1990. "The Real Objective of Mendel's Paper." *Biology and Philosophy* **5**:267–92.

Monod, J. 1971. *Chance and Necessity: An Essay on the Natural Philosophy of Modern Biology.* New York: Knopf.

Monod, J., and Jacob, F. 1961. "Genetic Regulatory Mechanisms in the Synthesis of Proteins." *Journal of Molecular Biology* **3**:318–56.

Monod, J., Changeux, J., and Jacob, F. 1963. "Allosteric Proteins and Cellular Control Systems." *Journal of Molecular Biology* **6**:306–29.

Monod, J., Wyman, J., and Changeoux, J. 1965. "On the Nature of Allosteric Transitions: A Plausible Model." *Journal of Molecular Biology* **12**:88–118.

Moran, P. A. P. 1973. "A Note on Heritability and the Correlation between Relatives." *Annals of Human Genetics* **37**:217.

Moran, P. A. P., and Smith, C. A. B. 1966. "Commentary on R. A. Fisher's Paper on the Correlation between Relatives on the Supposition of Mendelian Inheritance." *Eugenics Laboratory Memoirs* **41**:1–62.

Morgan, T. H. 1910. "Sex Limited Inheritance in Drosophila." *Science* **32**:120–2.

Morgan, T. H., Bridges, C., and Sturtevant, A. H. 1925. "The Genetics of Drosophila." *Bibliographia Genetica* **2**:1–262.

Morgan, T. H., Sturtevant, A. H., Muller, H. J., and Bridges, C. B. 1915. *The Mechanism of Mendelian Heredity.* New York: Henry Holt.

Muller, H. J. 1927. "Artificial Transmutation of the Gene." *Science* **66**:84–7.

Mullis, K., Falcoona, F., Scharf, S., Saiki, R., Horn, G., and Erlich, H. 1986. "Specific Enzymatic Amplification of DNA In Vitro: the Polymerase Chain Reaction." *Cold Spring Harbor Symposia in Quantitative Biology* **51**:263–73.

Murayama, M. 1966. "Molecular Mechanism of Red Cell 'Sickling'." *Science* **153**:145–149.

References

Nagel, E. 1949. "The Meaning of Reduction in the Natural Sciences." In Stauffer, R. C. (ed.), *Science and Civilization.* Madison: University of Wisconsin Press. pp. 99–135.

Nagel. E. 1961. *The Structure of Science.* New York: Harcourt, Brace, and World.

Nagylaki, T. 1992. *An Introduction to Theoretical Population Genetics.* Berlin: Springer-Verlag.

Needham, J., and Baldwin, E. (eds.) 1949. *Hopkins & Biochemistry: 1861–1947.* Cambridge, UK: W. Heffer and Sons.

Neurath, O. [1931] 1996. "Physicalism." In Sarkar, S. (ed.), *Science and Philosophy in the Twentieth Century: Basic Works of Logical Empiricism. Vol. 2. Logical Empiricism at Its Peak.* New York: Garland Publishing Inc. pp. 74–9.

Nickles, T. 1973. "Two Concepts of Inter-theoretic Reduction." *Journal of Philosophy* **70**:181–201.

Norton, B. J. 1975. "Biology and Philosophy: The Methodological Foundations of Biolmetry." *Journal of the History of Biology* **8**:85–93.

Olby, R. C. 1974. *The Path to the Double Helix.* Seattle: University of Washington Press.

Olby, R. C. 1979. "Mendel No Mendelian?" *History of Science* **17**:57–72.

Olby, R. C. 1985. *Origins of Mendelism.* Chicago: University of Chicago Press.

Oppenheim, P., and Putnam, H. 1958. "The Unity of Science as a Working Hypothesis." In Feigl, H., Scriven, M., and Maxwell, G. (eds.), *Concepts, Theories, and the Mind-Body Problem.* Minneapolis: University of Minnesota Press. pp. 3–36.

Ott, J. 1991. *Analysis of Human Genetic Linkage.* Baltimore: Johns Hopkins Press.

Owen, M. J., and Lamb, J. R. 1990. *Immune Recognition.* Oxford: IRL Press.

Painter, T. S. 1933. "A New Method for the Study of Chromosome Rearrangements and Plotting of Chromosome Maps." *Science* **78**:585–6.

Pato, C. N., Macciardi, F., Pato, M. T., Verga, M., and Kennedy, J. L. 1993. "Review of the Putative Association of Dopamine D2 and Alcoholism: A Meta-Analysis." *American Journal of Medical Genetics (Neuropsychiatric Genetics)* **48**:78–82.

Paul, D. B. 1984. "Eugenics and the Left." *Journal of the History of Ideas* **45**:567–90.

Paul, D. B. 1995. "Toward a Realistic Assessment of PKU Screening." In Hull, D. L., Forbes, M., and Burian, R. M. (eds.), *PSA 1994: Proceedings of the 1994 Biennial Meeting of the Philosophy of Science Association.* Vol. 2. East Lansing: Philosophy of Science Association. pp. 322–8.

Pauling, L., and Corey, R. B. 1950. "Two Hydrogen-Bonded Spiral Configurations of the Polypeptide Chains." *Journal of the American Chemical Society* **71**:5349.

Pauling, L., and Corey, R. B. 1951. "The Pleated Sheet, A New Layer Configuration of Polypeptide Chains." *Proceedings of the National Academy of Sciences (USA)* **37**:251–6.

Pauling, L., Itano, H. A., Singer, S. J., and Wells, I. C. 1949. "Sickle Cell Anemia, A Molecular Disease." *Science* **110**:543–8.

Pearson, K. 1900. *The Grammar of Science.* 2nd ed. London: A. and C. Black.

References

Pearson, K. 1893. "Contributions to the Mathematical Theory of Evolution." *Journal of the Royal Statistical Society* **56**:675–9.

Pearson, K. 1904a. "Mathematical Contributions to the Theory of Evolution. XII. On a Generalized Theory of Alternative Inheritance, with Special Reference to Mendel's Laws." *Philosophical Transactions of the Royal Society (London) A* 203:53–86.

Pearson, K. 1904b. "A Mendelian's View of the Law of Ancestral Inheritance." *Biometrika* **3**:109–12.

Perelson, A. S. 1989. "Immune Network Theory." *Immunological Reviews* **110**:5–36.

Perutz, M. F. 1951. "New X-Ray Evidence on the Configuration of Polypeptide Chains." *Nature* **167**:1053–4.

Perutz, M. F. 1990. *Mechanisms of Cooperativity and Allosteric Regulation in Proteins*. Cambridge, UK: Cambridge University Press.

Plomin, R. 1994. *Genetics and Experience: The Interplay Between Nature and Nurture*. Thousand Oaks, CA: Sage Publications.

Plomin, R. 1997. "Identifying Genes for Cognitive Abilities and Disabilities." In Sternberg, R. J., and Grigorenko, E. (eds.), *Intelligence, Heredity and the Environment*. Cambridge, UK: Cambridge University Press. pp. 89–104.

Plomin, R., DeFries, J. C., and McClearn, G. E. 1990. *Behavioral Genetics: A Primer. 2nd ed.* New York: W. H. Freeman.

Plomin, R., and Loehlin, J. C. 1989. "Direct and Indirect IQ Heritability Estimates: A Puzzle." *Behavior Genetics* **19**:331–42.

Plomin, R., Owen, M. J., and McGuffin, P. 1994. "The Genetic Basis of Complex Human Behaviors." *Science* **264**:1733–9.

Popper, K. 1957. "The Aim of Science." *Ratio* **1**:24–35.

Portin, P. 1993. "The Concept of the Gene: Short History and Present Status." *Quarterly Review of Biology* **56**:173–223.

Potter, H., and Dressler, D. 1978. "DNA Recombination: In Vivo and In Vitro Studies." *Cold Spring Harbor Symposia on Quantitative Biology* **43**:969–85.

Provine, W. B. 1971. *The Origins of Theoretical Population Genetics*. Chicago: University of Chicago Press.

Provine, W. B. 1992. "The R. A. Fisher-Sewall Wright Controversy." *Boston Studies in the Philosophy of Science* **142**:201–29.

Provine, W. B. 1986. *Sewall Wright and Evolutionary Biology*. Chicago: University of Chicago Press.

Quine, W. V. O. 1964. "Ontological Reduction and the World of Numbers." *Journal of Philosophy* **61**:209–16.

Quine, W. V. O. [1977] 1979. "Facts of the Matter." In Shahan, W., and Swoyer, C. (eds.), *Essays on the Philosophy of W. V. Quine*. Norman: University of Oklahoma Press. pp. 155–69.

Ramsey, J. L. 1995. "Reduction by Construction." *Philosophy of Science* **62**:1–20.

Richards, F. M. 1991. "The Protein Folding Problem." *Scientific American* **264**(1): 54–63.

References

Risch, N., and Botstein, D. 1996. "A Manic Depressive History." *Nature Genetics* **12**:351–3.

Rosenberg, A. 1978. "The Supervenience of Biological Concepts." *Philosophy of Science* **45**:368–86.

Rosenberg, A. 1985. *The Structure of Biological Science.* Cambridge, UK: Cambridge University Press.

Rosenberg, A. 1994. *Insturmental Biology or the Disunity of Science.* Chicago: University of Chicago Press.

Rotter, J. I. 1981. "The Modes of Inheritance of Insulin-Dependent Diabetes Mellitus or the Genetics of IDDM, No Longer a Nightmare but Still a Headache." *American Journal of Human Genetics* **33**:835–51.

Roughgarden, J. 1979. *Theory of Population Genetics and Evolutionary Ecology: An Introduction.* New York: Macmillan.

Ruse, M. 1971. "Reduction, Replacement, and Molecular Biology." *Dialectica* **25**:39–72.

Ruse, M. 1973. *The Philosophy of Biology.* London: Hutchinson.

Ruse, M. 1976. "Reduction in Genetics." *Boston Studies in the Philosophy of Science* **32**:631–51.

Rushton, J. P. 1995. *Race, Evolution, and Behavior: A Life History Perspective.* New Brunswick, NJ: Transaction.

Salmon, W. 1971. *Statistical Explanation and Statistical Relevance.* Pittsburgh, PA: University of Pittsburgh Press.

Salmon, W. 1984. *Scientific Explanation and the Causal Structure of the World.* Princeton, NJ: Princeton University Press.

Sapp, J. 1987. *Beyond the Gene: Cytoplasmic Inheritance and the Struggle for Authority in Genetics.* New York: Oxford University Press.

Sarkar, S. 1988. "Natural Selection, Hypercycles and the Origin of Life." In Fine, A., and Leplin, J. (eds.), *PSA 1988: Proceedings of the 1988 Biennial Meeting of the Philosophy of Science Association.* Vol. 1. East Lansing: Philosophy of Science Association. pp. 197–206.

Sarkar, S. 1989. "Reductionism and Molecular Biology: A Reappraisal." Ph.D. Dissertation, Department of Philosophy, University of Chicago.

Sarkar, S. 1991. "Reductionism and Functional Explanation in Molecular Biology." *Uroboros* **1**(1):67–94.

Sarkar, S. 1992. "Models of Reduction and Categories of Reductionism." *Synthese* **91**:167–94.

Sarkar, S. 1996. "Biological Information: A Skeptical Look at Some Central Dogmas of Molecular Biology." In Sarkar, S. (ed.), *The Philosophy and History of Molecular Biology: New Perspectives.* Dordrecht: Kluwer. pp. 187–231.

Sarkar, S. In press. "Physical Approximations and Stochastic Processes in Einstein's 1905 Paper on Brownian Motion." *Einstein Studies.*

Savageau, M. A. 1992. "Dominance According to Metabolic Control Analysis: Major Achievement or House of Cards?" *Journal of Theoretical Biology* **154**:131–6.

References

Schaffner, K. F. 1967a. "Antireductionism and Molecular Biology." *Science* **157**: 644–7.

Schaffner, K. F. 1967b. "Approaches to Reduction." *Philosophy of Science* **34**:137–47.

Schaffner, K. F. 1969. "The Watson-Crick Model and Reductionism." *British Journal for the Philosophy of Science* **20**:325–48.

Schaffner, K. F. 1974. "The Peripherality of Reductionism in the Development of Molecular Biology." *Journal of the History of Biology* **7**:111–39.

Schaffner, K. F. 1976. "Reduction in Biology: Prospects and Problems." *Boston Studies in the Philosophy of Science* **32**:613–32.

Schaffner, K. F. 1977. "Reduction, Reductionism, Values, and Progress in the Biomedical Sciences." In Colodny, R. (ed.) *Logic, Laws, and Life.* Pittsburgh, PA: University of Pittsburgh Press. pp. 143–71.

Schaffner, K. F. 1993a. *Discovery and Explanation in Biology and Medicine.* Chicago: University of Chicago Press.

Schaffner, K. F. 1993b. "Theory Structure, Reduction, and Disciplinary Integration in Biology." *Biology and Philosophy* **8**:319–47.

Schmalhausen, I. 1949. *Factors of Evolution: The Theory of Stabilizing Selection.* Philadelphia: Blakiston.

Schrödinger, E. 1944. *What is Life? The Physical Aspect of the Living Cell.* Cambridge, UK: Cambridge University Press.

Schwartz, R. J. 1977. "Discussion: Idealizations and Approximations in Physics." *Philosophy of Science* **45**:595–603.

Sherrington, R., Brynjolfsson, J., Petursson, H., Potter, M., Dudleston, K., Barraclough, B., Wasmuth, J., Dobbs, M., and Gurling, H. 1988. "Localization of a Susceptibility Locus for Schizophrenia on Chromosome 5. *Nature* **336**:164–7.

Shimony, A. 1987. "The Methodology of Synthesis: Parts and Wholes in Low-Energy Physics." In Kargon, R., and Achinstein, P. (eds.), *Kelvin's Baltimore Lectures and Modern Theoretical Physics.* Cambridge, MA: MIT Press. pp. 399–423.

Shimony, A. 1989. "Conceptual Foundations of Quantum Mechanics." In Davies, P. C. W. (ed.), *The New Physics.* Cambridge, UK: Cambridge University Press. pp. 373–95.

Shockley, W. 1972. "Dysgenics, Geneticity, Raceology: A Challenge to the Intellectual Responsibility of Educators." *Phi Beta Kappan* **53**:297–307.

Sklar, L. 1967. "Types of Inter-Theoretic Reduction." *British Journal for the Philosophy of Science* **18**:109–24.

Sklar, L. 1993. *Physics and Chance: Philosophical Issues in the Foundations of Statistical Mechanics.* New York: Cambridge University Press.

Smith, C. W., Patton, J. G., and Nadal-Ginard, B. 1989. "Alternative Splicing in the Control of Gene Expression." *Annual Review of Genetics* **23**:527–77.

Smith, K. C. 1992. "The New Problem of Genetics: A Response to Gifford." *Biology and Philosophy* **7**:331–48.

Sneed, J. 1971. *The Logical Structure of Mathematical Physics.* Dordrecht: Reidel.

References

Sober, E., and Wilson, D. S. 1994. "A Critical Review of Philosophical Work on the Units of Selection Problem." *Philosophy of Science* **61**:534–55.

Stegmüller, W. 1976. *The Structure and Dynamics of Theories.* Amsterdam: Springer-Verlag.

Stein, H. 1958. "Some Philosophical Aspects of Natural Science." Ph.D. Dissertation, Department of Philosophy, University of Chicago.

Stein, H. 1989. "Yes, But . . . Some Skeptical Remarks on Realism and Anti-Realism." *Dialectica* **43**:47–65.

Stent, G. S. 1968. "That Was the Molecular Biology that Was." *Science* **160**:390–5.

Stern, C. 1973. *Principles of Human Genetics.* 3rd ed. San Francisco: Freeman.

Sternberg, R. J., and Grigorenko, E. (eds.) 1997. *Intelligence, Heredity and the Environment.* Cambridge, UK: Cambridge University Press.

Stigler, S. M. 1986. *The History of Statistics: The Measurement of Uncertainty before 1900.* Cambridge, MA: Harvard University Press.

Strickberger, M. W. 1976. *Genetics.* 2nd ed. New York: Macmillan.

Stryer, L. 1988. *Biochemistry.* 3rd ed. New York: W. H. Freeman.

Sturdy, S. 1988. "Biology as Social Theory: John Scott Haldane and Physiological Regulation." *British Journal for the History of Science* **21**:315–40.

Sturtevant, A. H. 1913. "The Linear Arrangement of Six Sex-Linked Factors in Drosophila, as Shown by Their Mode of Association." *Journal of Experimental Zoology* **14**:43–59.

Suppes, P. 1957. *Introduction to Logic.* Princeton, NJ: Van Norstrand.

Sutton, W. S. 1903. "The Chromosomes in Heredity." *Biological Bulletin of the Marine Biological Laboratory, Woods Hole* **4**:231–48.

Swanson, J. W. 1962. "On the Kemeny-Oppenheim Treatment of Reduction." *Philosophical Studies* **13**:94–6.

Tanford, C. 1980. *The Hydrophobic Effect: Formation of Micelles and Biological Membranes.* 2nd ed. New York: Wiley.

Thaler, D. 1996. "Paradox as Path: Pattern as Map." In Sarkar, S. (ed.), *The Philosophy and History of Molecular Biology: New Perspectives.* Dordrecht: Kluwer. pp. 233–48.

Thompson, E. A. 1986. *Pedigree Analysis in Human Genetics.* Baltimore: Johns Hopkins Press.

Tinkle, W. J. 1927. "Heredity of Habitual Wandering." *Journal of Heredity* **18**:548–51.

Trow, A. H. 1913. "Forms of Reduplication – Primary and Secondary." *Journal of Genetics* **2**:313–24.

Tucker, W. H. 1994. *The Science and Politics of Racial Research.* Urbana: University of Illinois Press.

van Fraassen, B. 1980. *The Scientific Image.* New York: Oxford University Press.

Vassar, R. J., Potel, M. J., and Josephs, R. 1982. "Studies in the Fiber to Crystal Transition of Sickle Cell Hemoglobin in Acidic Polyethylene Glycol." *Journal of Molecular Biology* **157**:395–412.

References

Vogel, F., and Motuisky, A. G. 1986. *Human Genetics: Problems and Approaches.* 2nd ed. Berlin: Springer-Verlag.

Waddington, C. H. 1957. *The Strategy of the Genes: A Discussion of Some Aspects of Theoretical Biology.* New York: Macmillan.

Wahlsten, D. 1990. "Insensitivity of the Analysis of Variance to Heredity-Environment Interaction." *Behavioral and Brain Sciences* **13**:109–20.

Waters, C. K. 1990. "Why the Anti-Reductionist Consensus Won't Survive the Case of Classical Mendelian Genetics." In Fine, A., Forbes, M., and Wessels, L. (eds.), *PSA 1990: Proceedings of the 1990 Biennial Meeting of the Philosophy of Science Association.* Vol. 1. East Lansing: Philosophy of Science Association. pp. 125–39.

Watson, J. D., and Crick, F. H. C. 1953a. "Genetical Implications of the Structure of Deoxyribonucleic Acid." *Nature* **171**:964–7.

Watson, J. D., and Crick, F. H. C. 1953b. "Molecular Structure of Nucleic Acids – A Structure for Deoxyribose Nucleic Acid." *Nature* **171**:737–8.

Watson, J. D., Tooze, J., and Kurtz, D. T. 1983. *Recombinant DNA: A Short Course.* New York: W. H. Freeman and Company.

Weatherall, D. J. 1991. *The New Genetics and Clinical Practice.* 3rd ed. Oxford: Oxford University press.

Weinberg, W. 1908. "Über den Nachweis der Vererbung beim Menschen." *Jahreshefte des Vereins für Vaterländische Naturkunde in Württemburg* **64**:368–82.

Weismann, A. 1889. *Essays upon Heredity and Kindred Subjects.* Oxford: Clarendon.

Wellems, T. E., and Josephs, R. 1980. "Helical Crystals of Sickle Cell Hemoglobin." *Journal of Molecular Biology* **137**:443–50.

Wellems, T. E., Vassar, R. J., and Josephs, R. 1981. "Polymorphic Assemblies of Double Strands of Sickle Cell Hemoglobin." *Journal of Molecular Biology* **153**:1011–26.

Weyl, H. 1949. *Philosophy of Mathematics and Natural Science.* Princeton, NJ: Princeton University Press.

Wilkins, M. H. F., Stokes, A. R., and Wilson, H. R. 1953. "Molecular Structure of Deoxypentose Nucleic Acids." *Nature* **171**:738–40.

Wilson, E. O. 1975. *Sociobiology: The New Synthesis.* Cambridge, MA: Harvard University Press.

Wimsatt, W. C. 1972. "Teleology and the Logical Structure of Function Statements." *Studies in the History and Philosophy of Science* **3**:1–80.

Wimsatt, W. C. 1976a. "Reductionism, Levels of Organization and the Mind-Body Problem." In Globus, G., Savodnik, I., and Maxwell, G. (eds.), *Consciousness and the Brain.* New York: Plenum. pp. 199–267.

Wimsatt, W. C. 1976b. "Reductive Explanation: A Functional Account." *Boston Studies in the Philosophy of Science* **32**:671–710.

Wimsatt, W. C. 1979. "Reduction and Reductionism." In Asquith, P. D., and Kyburg, H. (eds.), *Current Research in the Philosophy of Science.* East Lansing: Philosophy of Science Association. pp. 352–77.

References

Wimsatt, W. C. 1981. "Robustness, Reliability and Overdetermination." In Brewer, M., and Collins, B. (eds.), *Scientific Inquiry and the Social Sciences.* San Francisco: Jossey-Bass. pp. 124–63.

Wimsatt, W. C. 1992. "Golden Generalities and Co-Opted Anomalies: Haldane vs. Muller and the Drosophila Group on the Theory and Practice of Linkage Mapping." In Sarkar, S. (ed.), *The Founders of Evolutionary Genetics: A Centenary Reappraisal.* Dordrecht: Kluwer. pp. 107–66.

Wimsatt, W. C. 1995. "The Ontology of Complex Systems: Levels of Organization, Perspectives, and Causal Thickets." *Canadian Journal of Philosophy* **20**:207–74.

Woltereck, R. 1909. "Weitere experimentelle Untersuchungen über Artveränderung, speziell über das Wesen quantitativer Artunterschiede bei Daphnien." *Verhandlungen der Deutschen Zoologischen Gesellschaft* **19**:110–73.

Woodger, J. H. 1937. *The Axiomatic Method in Biology.* Cambridge, UK: Cambridge University Press.

Woodger, J. H. 1952. *Biology and Language.* Cambridge, UK: Cambridge University Press.

Wright, S. 1920. "The Relative Importance of Heredity and Environment in Determining the Piebald Pattern of Guinea Pigs." *Proceedings of the National Academy of Sciences (USA)* **6**:320–32.

Wright, S. 1921. "Correlation and Causation." *Journal of Agricultural Research* **20**:557–85.

Wright, S. 1929a. "The Evolution of Dominance." *American Naturalist* **63**:556–61.

Wright, S. 1929b. "Fisher's Theory of Dominance." *American Naturalist* **63**:274–9.

Wright, S. 1931. "Evolution in Mendelian Populations." *Genetics* **16**:97–159.

Yule, G. U. 1902. "Mendel's Laws and their Probable Relations to Intra-Racial Heredity." *New Phytologist* **1**(9):193–207, 222–38.

Yule, G. U. 1906. "On the Theory of Inheritance of Quantitative Compound Characters on the Basis of Mendel's Laws – A Preliminary Note." In Wilks, W. (ed.), *Report of the Third International Conference 1906 on Genetics.* London: Royal Horticultural Society. pp. 140–2.

Zallen, D. T. 1996. "Redrawing the Boundaries of Molecular Biology: The Case of Photosynthesis." In Sarkar, S. (ed.), *The Philosophy and History of Molecular Biology: New Perspectives.* Dordrecht: Kluwer. pp. 47–65.

Index

Index

cultural inheritance, 123–124, 127
cybernetics, 140, 201
cytoplasmic inheritance, 177–178

Darden, L., 205, 212
Davenport, C., 112
Davidson, D., 37
Dawe, C. M., 17, 28, 198
Delbrvck, M., 22, 64, 66, 205
determinism, 10–12, 37, 201
determinism, genetic, 10–12, 181
derivation, 42
developmental biology, 142, 153, 173
diabetes mellitus, 125
dialectical materialism, 63, 214
DNA double helix model, 2, 42, 138, 161
dominance, 72, 76–77, 96, 111, 152, 156,
   168–173, 219–220
dopamine D2 receptor, 3
Duchenne muscular dystrophy, 122

Earman, J., 193
Einstein, A., 203
eliminativism, epistemological, 60–62
eliminativism, ontological, 62–64
empiricism, logical, 16–19, 33, 64, 68, 69,
   151, 194–195, 217
epistasis, 72, 76–77, 96, 98, 219
epistemology-ontology distinction, 15, 17,
   20–23, 26, 37, 195, 197
eugenics, 1, 2, 191, 209
evolutionary theory, 47, 69, 140
explanation, 9–10, 19–20, 23–24, 29–32,
   39–43, 46–47, 52, 59, 64, 65, 68–70, 130,
   134, 140, 155, 161, 195, 197, 201–203,
   218–219
explanation, deductive-nomological model,
   9, 25–26, 29–30, 38, 40, 42, 48,
   196–200
explanation, functional, 69, 168–169,
   205, 219
explanation, historical, 47
explanation, Salmon model, 9, 29–32,
   37–38, 40–42, 193
explanation, statistical, 9, 102, 110–111,
   124, 197–198, 210
explanation, statistical relevance model,
   see explanation, Salmon's model
explanation, topological, 173
explanatory weight, 42, 62, 85, 140
explication, 185–187, 221
expressivity, 123, 124, 126, 128, 131–132,
   152, 177, 182, 184, 186, 188, 219
eye color, 8

F-justification, 47, 49–52, 62, 144, 202
F-realm, 47, 50, 52, 53, 59, 60, 62–65, 127,
   136–137, 147, 151, 202, 203, 205
F-rules, 47, 49, 50, 52, 53, 59, 62, 109,
   149–150, 156, 167, 169–170, 173
fact-convention distinction, 33
feedback, 142, 145
Feferman, S., 198
Feigl, H., 68
Feyerabend, P., 26–27
Fisher, R. A., 49, 71, 72, 82, 94, 105–109,
   124, 170, 185, 208, 211
Fodor, J., 37, 69, 154, 195, 196, 200
formal-substantive distinction, 9, 17, 18–20,
   24, 29, 39, 40, 199
Franklin, R., 138
Frege, G., 18
fundamentalism, epistemological, 26, 43,
   45–47, 52, 55, 63, 80–81

Galton, F., 52, 71, 105, 107, 211
Garrod, A., 1, 112
gene, cryptic, 157–158
gene, definition of, 156–159
gene, movable, 156–157
gene, overlapping, 157, 213
gene, repeated, 158
gene conversion, 162
gene editing, 157–158
gene expression, 139, 153, 159
gene mapping, 66
gene recombination, 6, 116, 159–161, 166
gene regulation, 56–59
gene replication, 139, 153, 159–166
gene splicing, 152
gene transmission, 152–153, 159–160, 162
genetic code, 2, 139, 152, 157–159
genetic drift, 134
genetic heterogeneity, 123, 126, 214
genetic network, 173
genetics, classical, see Mendelian genetics
genetics, human, 1, 2, 66, 92, 208–209
genetics, Mendelian, 6–7, 14, 34–35, 38, 42,
   44, 48, 49, 53, 61, 62, 71, 72, 96,
   101–135, 137, 139–140, 150–174,
   182–190, 192, 210–214, 215
genetics, molecular, 6, 38, 49, 61, 137,
   139–140, 150–174, 181, 182–190, 220
genetics, population, 6, 13, 49, 50, 52, 105,
   134–135, 211
genetics, quantitative, 13, 52, 71, 72,
   102, 211
Gifford, F., 192
Gilbert, S., 216

243

Index

Index

Index